Douglas H. Crawford

Dec 1990.

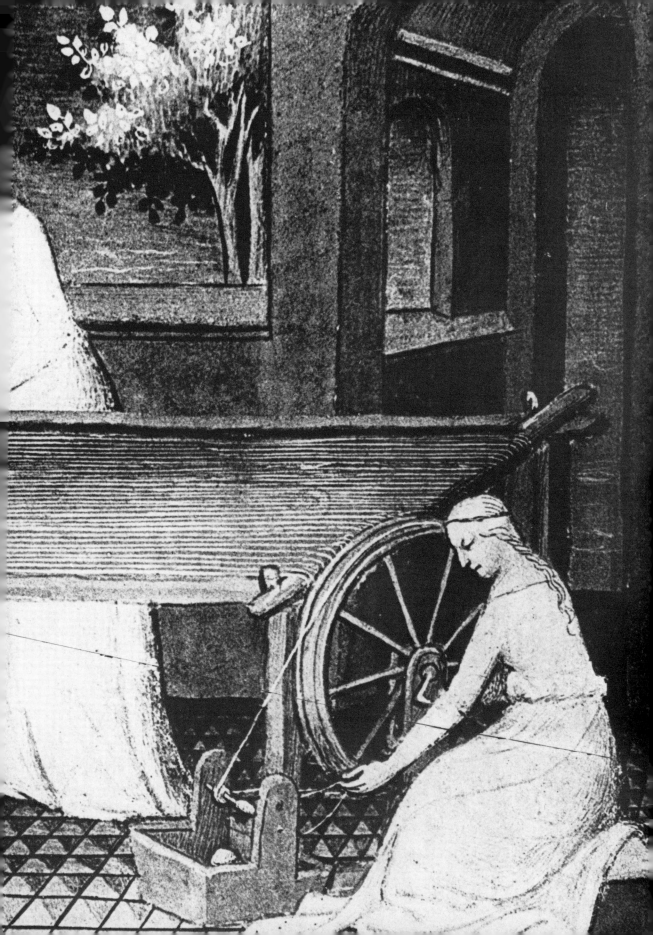

SCIENCE AND THE HOME

500 Years of technical advance

SCIENCE AND THE HOME

500 Years of technical advance

DOREEN YARWOOD

Science Consultant: Professor John Yarwood M.Sc., F.Inst.P.

B.T. Batsford Ltd.   London

ISBN 0 7134 5749 X

Filmset by Servis Filmsetting Ltd, Manchester
and printed in Great Britain by
The Anchor Brendon Ltd
Tiptree, Essex
for the publishers
B.T. Batsford Ltd
4 Fitzhardinge Street
London W1H 0AH

Contents

Preface

Since the Second World War it has been accepted in both school and college education that students studying science should also acquire a background knowledge of arts and language: such supplementary courses are generally described as 'liberal studies'. More recently the view has been put forward that students of history and allied subjects should be given some help in understanding the vital part which science and technology have played, and are increasingly playing, in determining the course of events which have shaped people's lives. This view is now expressed in the syllabuses for a number of the examinations set for the General Certificate of Education. Students taking subjects such as Social and Economic History and Home Economics are expected, for example, to study the effects of the Industrial Revolution and its modern technological counterpart, the development of new sources of power, the emergence of forms of transport and communication, of food science, synthetic materials, the rise of the chemical industry and mass entertainment by radio and television.

The idea of teaching an understanding of such subjects to students of history and domestic science is not new, but its implementation has been long delayed chiefly because it is not easy to explain much of the development of scientific and technological achievement to students who possess little if any scientific knowledge or mathematical ability. There are now available several popular books on science and television programmes for schools have pointed the way towards methods of explaining simply in words and pictures many of these concepts and happenings.

This book is an attempt to trace the way in which science and technology have changed domestic life since about AD 1500. It is written by an art historian for students studying Social and Economic History as well as Home Economics at 'O'-level standard. The close co-operation and generous assistance given to me by my husband John Yarwood have ensured its scientific accuracy. The introductory chapter discusses the outline effects of science and technology over these years and relates the contribution of each. The second chapter is also introductory; it deals chronologically with the emergence of sources of power (upon which all other technological development has depended) from slave labour to nuclear power stations. There are then five chapters on the basics of home life – food, shelter, clothes – three on the home itself, its structure, materials and interior, one on food and one on textiles. To these fundamentals I have added a chapter on home communication and entertainment. Finally there is a glance at the present microelectronics revolution and its possible meaning for the immediate domestic future. A selected bibliography is provided for students who wish to extend their studies beyond this brief introduction to the subject.

The author and publisher would like to thank the following for permission to reproduce the photographs in this book: W.J.H. Alexander Ltd (11); Architectural Press Ltd (125); The British Architectural Library, Royal Institute of British Architects (119); British Broadcasting Corporation (263); British Petroleum Company Ltd (26); British Telecom (234); Castle Farm Folk Museum, Marshfield (66); Central Electricity Generating Board (44); Cheltenham City Museum (80); Ercol Furniture Ltd (183); Fibreglass Ltd (93, 128); Findus Ltd (60); Flymo Ltd (131); Formica Ltd (136); Gas Council (72); Glass Manufacturers' Federation (92, 94, 127); Goblin BVC Ltd (38); ICI (95, 96, 181); Institute of Agricultural History and Museum of English Rural Life, University of Reading (12, 46, 48, 50–3, 109); Massey-Ferguson Ltd (54, 55); Mullard (255, 256); Museum of London (77); National Coal Board (17); Mrs Jean Naylor (39); Philips Electronic and Associated Industries Ltd (9, 31, 260); Philips Video (252); Pilkington Brothers Ltd (110); Science Museum, London (endpapers, 1–8, 10, 13, 14–16, 18–21, 23, 25, 49, 63, 75, 76, 79, 81, 83, 85, 86, 89, 90, 97, 115, 120, 129, 149, 157, 166, 170, 171, 185, 190, 195, 201, 205–8, 211, 222, 225, 227, 232, 236, 239, 241, 242, 243, 245); Science Museum/Crown Copyright (24, 36, 59, 69, 70, 73, 74, 140, 141, 143–5, 146–8, 153, 155, 159, 188, 199, 223, 224, 226, 228–30, 233, 235, 238, 244, 246–50); Science Museum, photographs by John Yarwood (142, 152, 156, 158, 200, 231, 237); lent to the Science Museum by the British Northrop Loom Co, Blackburn (206); lent to the Science Museum by Bryant and May (87); lent to the Science Museum by Electrical and Musical Instruments Ltd (251); lent to the Science Museum by Imperial College of Science and Technology (88); Singer Company (UK) Ltd (262); Sony (UK) Ltd (240, 253); Stranger's Hall Museum, Norwich (67); Thorn Domestic Appliances (Electrical) Ltd (71, 254); TI Creda Ltd (258, 259); TI Russell Hobbs Ltd (261); Tube Investments Ltd (37); Victoria and Albert Museum (78, 99, 100, 103, 132–5, 138, 139, 150, 178, 179, 180, 182); Wates Ltd (122, 126); Yorkshire Museum, York (137).

Doreen Yarwood
East Grinstead 1982

CHAPTER ONE

Science and Society

Both science and technology have been vital factors in moulding human society. Until comparatively recent times they have performed different, separate functions, the progress of one, more often than not, unrelated to that of the other. The English word science derives from the Latin *scientia*, meaning knowledge. In its modern sense it may be more closely defined as knowledge gained by observation, tested by experimentation, of the workings of nature. This concept of science, both intellectual and practical in approach, stems, in Britain, from the ideas of a small group of men in the seventeenth century who attempted to investigate the natural laws which govern the behaviour of matter. They studied the natural sciences, that is, biology, chemistry, physics.

Whereas science is knowledge of how and of what the world, indeed the universe, is made and operates, technology is, or should be, a method adopted by mankind to make life in this world easier, more comfortable and more pleasant by improving the quality of satisfying basic material needs such as shelter, food and clothing. The term technology derives from the Greek word *tekhnologia*, its usual dictionary definition being 'science of the industrial arts'. The modern interpretation tends to be rather wider than this.

For much of man's history since prehistoric times the improvements in his way of life resulted from an empirical approach, that is, a process of trial and error, by which means simple ways were developed of lifting or moving things,

harnessing natural forces to aid this and making equipment and tools to satisfy important needs. From such empirical methods technology has evolved. With each step forward further invention and mechanization has resulted.

The scientific revolution of the seventeenth century led, over 100 years later, to a technology which began to rely upon science. The growing wealth of scientific knowledge was not yet a replacement for the empirical system but it helped the technical innovator to see which path of experimentation might be more fruitful for his purpose. With the industrialization of the nineteenth century the bond between science and technology strengthened. In our own time the mutual reliance of one discipline upon the other has greatly increased. For example, technology is in debt to scientific research for knowledge of materials and energy resources, and also for the perception of new fields for study. Science appeals to technology to produce the instruments and equipment which it needs for research as well as for the handling and dissemination of data.

Yet, over the last 300 years technological innovation and scientific discovery have, most often, not marched in step. In some cases scientific research brought new knowledge but it was a long time, often more than a century, before it became possible to apply such knowledge technologically to produce something useful. For instance, Michael Faraday discovered electromagnetic induction in his laboratory experiments in 1831 but it was the 1880s before this understanding was translated into efficient electric generators and motors (page 39). An example for more recent times is the laser (where the delay between the scientific discovery and the techonological application is getting much shorter). Its action was first observed by Schawlow and Townes in 1958 but there was at that time no specific plan for its use; indeed, it was on occasion referred to as 'a solution looking for a problem'. Within a short while, however, its possibilities were perceived and utilized. It is now extensively employed in various fields, from opthalmic surgery to printing (page 173).

On the other hand innovatory ideas have been put forward and mechanisms manufactured to work perfectly well without the scientific principles behind such equipment being understood. This was particularly so prior to the seventeenth and eighteenth centuries (but to some extent is still true today) when men developed and designed artefacts on a trial and error basis. An example is the water-wheel, used as an instrument of power (page 24) in Britain from the early Middle Ages but not developed to its most efficient potential until John Smeaton carried out his scientific experiments in the eighteenth century to demonstrate the superior efficacy of the overshot design. Also, over the years, technological ideas have been advanced but, due to the inadequate state of current scientific knowledge, were stillborn. In some instances such ideas have been put into cold storage, then resuscitated at a later time when conditions were more favourable. A number of proposals were put forward early for uses to which semiconductors might be put. Also, the first radio receiver made use of a crystal (page 162). Yet it was not until the 1950s that chemical and metallurgical techniques, among others, had been developed to such a sophisticated level that the semiconductor diode and the transistor were commercially realized (pages 164, 169).

It is sometimes suggested that many scientific discoveries are accidental and that a number of the outstanding technological ideas were chanced upon. It is true that some innovations derived from lines of research which were not in the mainstream of the work upon which the scientist or inventor was currently engaged and that an element of chance or coincidence was present. A notable recent example is the discovery of penicillin by Alexander Fleming in 1928. But the majority of discoveries and vital developments were due to repeated study and trial and error and only a notable, analytical mind, ready to interpret and grasp each new factor, would be critically aware and so take advantage of such chance. Louis Pasteur's comment that 'Chance favours only the mind which is prepared' is apt.

In science, as in the arts, new discoveries and conclusions were, in many instances, reached almost simultaneously by people of different nationalities working independently and often unknown to one another. It seems as if knowledge in any particular sphere reaches a certain stage of gestation so that the time is ripe for the next venture to be made. Of the numerous instances of this may be mentioned the various contenders put forward for the invention in the fifteenth century of typographical printing: Gutenberg. Fust, Schoeffer, Laurens Coster (page 154), or the nineteenth-century development of the incandescent filament electric lamp by the American Thomas Edison and the Englishman Sir Joseph Swan (page 119).

Over the centuries scientific knowledge has played a comparatively small part in the development of some industries: steel, for example, and the early stages of photography. In others, particularly those of the nineteenth and twentieth centuries, its contribution has been vital and extensive as it was in the chemical and electrical industries. In our own century the time lag between scientific discovery and technical development has become steadily shorter and the dependence of one discipline upon the other greater. This is particularly so in communications (for instance, television) and electronics.

THE EARLY YEARS: EMPIRICISM

From the earliest times until the Renaissance all advances were made by empirical means, that is, by way of experimentation. The word empiric derives from the Greek *empeirikos*, meaning experience. Fundamental discoveries, such as that by primitive man of how to make fire were followed in the third millennium BC by the development of the wheel. After solving the means to cook food, the greatest problem in early centuries was the creation of power (Chapter Two) and the means of transportation.

The first stages in developing these functions was man's ability to domesticate cattle. It was soon appreciated that a castrated animal became biddable and the breeding of heavy draught animals was undertaken. Such animals were trained to pull heavy wooden carts with solid wheels. It was natural that this development should take place in areas where the land was flat, the earth firm yet where timber was readily available for construction purposes. Several parts of Europe and Asia were suitable and such wheeled transport appeared in Mesopotamia, southern Russia, Armenia, Georgia and along the Baltic coastal regions from Holland to Poland.

The classical world of Greece and Rome developed all kinds of tools and equipment as well as acquiring engineering and building skills. Slaves provided a basic source of power but the Christian populations of the Middle Ages had no such means available to them and much of their energies were devoted to developing ways of providing power for pumping, hauling and grinding. They harnessed water and air to turn water-wheels and windmills for their basic needs of food, shelter and clothes and dug shallowly in the earth to mine metals and chemicals.

THE SCIENTIFIC REVOLUTION

In the two centuries between 1500 and 1700 the basis of life was irrevocably altered for everyone in Europe, though at first this was only perceived by the small educated minority who were able to read and understand the exciting new ideas about the known world which were being circulated. The Scientific Revolution, as it is now termed, was essentially related to and a part of the wider Renaissance movement which had begun in literary form in fourteenth-century Italy with the writings of Petrarch, Boccaccio and Dante, then gradually gaining momentum with its extension into the media of the visual arts and architecture. The concepts of the Renaissance spread westward and northward until, by the sixteenth century, they dominated the thinking of artists and writers in the whole of western

Europe. It was then that the new methods of thinking, of approach, began to extend to science.

A number of different reasons have been put forward why the Scientific Revolution should have taken place in the sixteenth and seventeenth centuries in Europe. These include the extension of the world known to Europeans by means of geographical exploration, economic advances leading to greater wealth and higher standard of living, the advent of printing which enabled new knowledge to be disseminated and, not least, the existence, at that time, of a few men of outstanding genius living contemporaneously in several European countries. Certainly all these factors contributed to the establishment then of the new scientific thought but without the invention of typographical printing it would have been a markedly slower process (page 154).

In the Middle Ages knowledge was disseminated with difficulty by means of manuscript copies. By 1500 printing presses were available in nearly every European country, better paper-making methods were being experimented with and a flood of books and papers, scientific works among them, were published and read; the printing industry expanded rapidly during both sixteenth and seventeenth centuries (1). The revived interest in the ancient classical world applied to science as well as literature and architecture and the works and ideas of philosophers and scientists, for example, Aristotle and Archimedes, were studied by a wider public. At the same time the Greek theories were being challenged; Aristotle's concept of the universe and, in particular, his view of the laws of motion, were replaced as a result of the modern scientific method of observation and experimentation taken up by scientists such as Galileo and Newton.

The global exploration which had been widely undertaken in the fifteenth and sixteenth centuries had extended the scientific understanding of the world. It also demanded more advanced navigational techniques (e.g. the magnetic compass resulting from the pioneer work of the Elizabethan scientist, Gilbert) and improved maps and charts. Their exploration also brought wealth to the European countries of origin and

1   *Printing in 1574 (page 154)*

this, together with advances in agricultural methods, led to an improvement in economic standards.

All these factors – printing, exploration, improved economic levels – brought about a climate suitable for the creation of a new intellectual approach to science. It was the genius of the individual scientists who established a novel tradition and made the experiments and observations leading to the fundamental discoveries which laid the foundations of nineteenth- and twentieth-century science and technology. This was the work of such men as Galileo and Leonardo in Italy, Descartes in France, Copernicus in Poland, Tycho Brahe in Denmark, Kepler in Germany and Newton in England.

The scientific method of investigation developed in the seventeeth century was found to require a more reliable and sophisticated system of mathematics as well as a greater variety of instruments of improved accuracy. Impressive advances were made during the century in both these areas, by and large mainly by the scientists who themselves were engaged upon the important discoveries of the day in the fields of astronomy and mechanics. By the end of the century, for instance, the simple mathematical symbols for addition, subtraction, multiplication and division had come into general use and logarithms had been invented by the Scottish mathematician John Napier (1550–1617). This technique saved time and effort in calculation by reducing the need to multiply and divide to addition and subtraction. More important still was the invention of the calculus, one type by Isaac Newton (1642–1727) in England and another by Gottfried von Leibniz (1646–1716) in Germany, each working independently of the other. Of considerable usefulness as a calculating device was the introduction of the graduated rule (the sector); this was followed later in the century by the slide rule.

Demand for scientific instruments of quality and accuracy grew rapidly during the seventeenth century as the new knowledge was more widely disseminated. A number of scientists, chiefly astronomers such as Tycho Brahe (1546–1601) had designed and made the instruments which they needed. Craftsmen were then trained to copy and make numbers of such instruments. As the craft of instrument-making grew, the demand increased further with the printing of yet more books (page 154). Instruments included navigational aids such as the quadrant and magnetic compass, measuring devices such as the thermometer and barometer and optical instruments, for example, the telescope and microscope.

In 1500 engineering techniques were still entirely empirical, but during the sixteenth and seventeenth centuries a closer relationship developed between the scientific investigation of mechanics and dynamics and an engineering method more closely founded upon the results of such investigations. Precursor of this trend was that unique Renaissance genius Leonardo da Vinci (1452–1519), artist, philosopher, scientist and engineer, who filled so many notebooks with drawings which antedated a tremendous variety of concepts and mechanisms from spinning wheels and screw-cutting lathes to tanks and flying machines. A century before Francis Bacon

By the mid-seventeenth century knowledge of science and technology was not only being spread by books but also, increasingly, by correspondence and discussion. Scientists in several countries began to meet regularly to talk together about the new experimental philosophy. From such meetings were formed, notably in England, France and Italy, professional societies dedicated to a dissemination of the results of research by means of scientific papers and journals. Committees were set up to survey the application of industrial technology to different crafts and trades and new machines and inventions were carefully evaluated and reported upon.

The English society was founded in 1660, just after the Restoration, and was called a *Society of Experimental Philosophy*. Two years later it was granted a royal charter by Charles II as *The Royal Society of London for Improving Natural Knowledge*; it has been known as *The Royal Society* ever since. One of the chief differences between it and its sister organisation in Paris, the *Académie Royale des Sciences* (founded 1666), and its lasting strength, has been its total independence from government control.

Although the principal disciplines studied by the Royal Society were science and philosophy, founder and early members included men of academic stature from related walks of life such as Christopher Wren and John Evelyn. They also included the physicists Robert Boyle and Robert Hooke but the dominating influence later in the seventeenth century was that of England's great scientist Isaac Newton (4) (1642–1727), who was its president in 1703 and when he died at the age of 84, became the only English scientist to be buried in Westminster Abbey. Newton's outstanding contribution to scientific understanding and knowledge encompassed a broad field in natural philosophy and, though he owed a considerable debt to his predecessors – Galileo, Descartes, Kepler, for example – his clear enunciation of the laws of motion and his studies of gravitation and optics provided for his successors a completely new basis on which to work.

2   *Robert Boyle (1627–91). Portrait engraving*

put forward his theories upon the importance of experimentation in scientific method, Leonardo was advocating this. 'Sciences are vain and full of errors which are not born from experiment', he said. Many of the ideas which Leonardo put forward were tested out and further investigated in the seventeeth century and these led to practical engineering results in the industrial revolution 100 years later.

The studies of Galileo, Torricelli, von Guericke (3) and Boyle (2) into the properties of air, of atmospheric pressure and vacuum led directly, in the eighteenth century, to the development of the atmospheric steam engine (pages 28–32). Navigational requirements of maritime nations, notably Britain and Holland, resulted in the achievement of more accurate timekeeping. The introduction of the pendulum by the Dutchman Huygens in 1657, which was a fundamental improvement, owed much to the experimental study of dynamics undertaken earlier by Galileo (1564–1642).

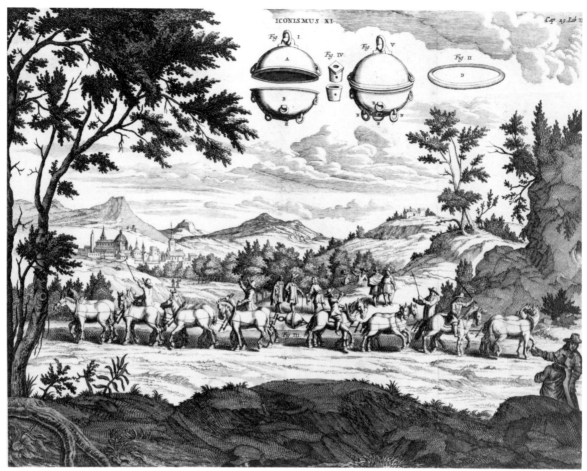

3   *Print of Otto von Guericke's experiment on atmospheric pressure at Magdeburg in 1657 with two copper hemispheres and 16 horses. The hemispheres are now in* the Deutsches Museum, Munich *(page 28)*

THE INDUSTRIAL REVOLUTION AND ITS SEQUEL (1750–1918)

The term 'Industrial Revolution' has long been used by historians to describe the events of the eighteenth and nineteenth centuries which, in a mere three generations, changed the face of Britain. In 1750 the country was still agricultural, that is, over 80 per cent of the population were dependent on the land for their livelihood and only a small percentage worked in towns. A hundred years later half the population still lived in rural areas but only 22 per cent worked on the land. And, even in such rural areas, life was completely changed as the land was gradually enclosed (page 46) and was increasingly bisected by turnpike roads, canals and, later, railways.

Some modern historians question, though, whether the phenomenon should be described as a revolution or whether it was really evolution at an accelerated pace. Others maintain that the movement contained a revolutionary element because totally new factors were introduced into the evolutionary process. Two such factors were the gradual substitution of iron and steel for wood as a structural material and the development of the factory system of manufacture by means of the invention of machinery to replace the hand-working methods. In fact, both these factors depended heavily for their emergence upon a more fundamental new element: the introduction of steam power (page 27). Without the steam engine the production of coal could not have been so greatly increased and the development of iron- and steel-making so advanced in

4　*Sir Isaac Newton, from a painting by Kneller c.1712*

order to manufacture so much machinery.

Yet, modern historical research has shown that the industrialization of the time was to a large extent of an evolutionary nature since a number of the factors which characterize the Industrial Revolution were present in Britain long before 1750 and so did not reach their full development until the later nineteenth century. For example, the ideas of standardization of machine parts, the use of factory techniques and rationalization of the use of labour had been well understood in the early eighteenth century. Also, water and wind power had been made to function efficiently; water-wheels, in particular, were widely used in, for example, textile and corn mills. But it was the introduction and perfecting of steam power which paved the way to extensive industrialization and technical development and it was these which made possible the rise of modern civilization.

That it was Britain which was the first country to experience industrialization, which was the birthplace of so many of the inventions and technical achievements making this possible and which, by the early years of the nineteenth

century, had become known as the 'workshop of the world', is undisputed. That this was so was not a chance happening. There were many contributory reasons of which the most important were:

1　In the years 1750–1850 the countries of Europe were torn apart by periodic wars but Britain was much less involved than other nations.
2　The Act of Union of 1707 had made the whole of Britain one economic unit.
3　The country had established a religious tolerance not yet achieved by a number of other European countries and so was able to welcome skilled refugee workers to British shores – the Huguenots, for example.
4　The country possessed ample reserves of coal and iron ore.
5　As an island Britain had long been a trading maritime nation and so was able to export the excess merchandise produced by the factory system of production.

Due to these factors and many more, the inventive abilities and business acumen of the British people were able to exploit the opportunities of the time.

Paradoxically one of the factors which enabled the Industrial Revolution to succeed was the same one which largely created the overcrowded housing conditions in towns. This was an unprecedentedly rapid growth of population. In 1750 the population of England and Wales was about six-and-a-half million; a century later the figure had reached almost 18 million. The phenomenal rise, added to a thriving export market, enabled manufacturers to sell comfortably the greatly increased quantity of goods – textiles, ceramics, building materials, food, etc. – which were being produced by mechanized factory methods and so to create a further extension in production.

There were several major reasons for this rapid population increase which continued unabated during the nineteenth century, reaching 36 million in England and Wales by 1911, but they did not include a notable rise in the birth-rate. Also, until the 1840s, when starvation due to the failure of successive potato crops brought a flood of Irish into England, there had been no appreciable rise in population due to immigration. The population figures rose so dramatically because of a steep and continuing fall in the death-rate. Many factors contributed to this; for example, improved agricultural methods led to better feeding, the heavy tax on gin brought to an end scenes like that portrayed by Hogarth (Gin Lane), standards of hygiene, medical care and personal cleanliness all improved, towns were paved and sewage and water provisions were radically overhauled (page 96).

The contribution of science and technology to the development of industrialization gained momentum during the later eighteenth and the nineteenth centuries. Empirical methods continued to be followed, certainly until the later nineteenth century. James Watt, for instance, adopted trial and error techniques for some time in improving the design of his steam engine (page 30). Edison worked his way through countless experiments until he found a satisfactory answer to his investigations in many fields, the incan-

descent filament electric lamp, for instance (page 119) and the phonograph (page 157). Even Bessemer, it is said, did not fully comprehend the chemical actions taking place in his converter, his knowledge of chemistry being elementary (page 67).

Yet, despite this continued empiricism, the modern scientific approach, based on the example and mathematical work of Sir Isaac Newton and his fellow seventeenth-century scientists, gained adherents and gradual acceptance. The French chemist, Antoine Lavoisier (1743–94), laid the foundations of modern chemistry with his quantitative methods of chemical investigation. Many scientists from Britain and Europe contributed to a gradual, more comprehensive understanding of the nature of matter and the changes occurring in its solid, liquid and gaseous states (page 27). Oxygen was discovered and identified and hydrogen, also new metals such as potassium and sodium. The English chemist and physicist, John Dalton (1766–1844), propounded his atomic theory (1801) in which he

5   *Factory use of mechanical textile processes, 1835. Carding, drawing and roving (page 135)*

6   *Steam-powered Aveling and Porter reaper, 1876 (from* Implement and Machinery Review*) page 50)*

suggested a basic chemical concept that matter consists of tiny indivisible atoms*, that in an element all atoms are identical and that atoms from different elements have different weights.

During the second half of the nineteenth century scientific knowledge became more important to technological progress though the delay between concept and realization was still often considerable. A marriage of the two disciplines led to notable improvements in the standard of living and quality of life of a large part of the population (5, 6). For instance, increased yields and more nutritious foods resulted from studies of agricultural methods, soil composition and fertilizers; greater knowledge led to better medical care and a reduction in disease (page 46–7); the discoveries and inventions in the realms of fuel and power brought gas and electric lighting and heating as well as powered machines to perform laborious tasks (Chapter Two); studies of electricity and magnetism, as well as acoustics, ushered in such wonders as the telegraph, the telephone (7) and the phonograph (8) (page 157).

*We now know that they are fissionable (page 39).

One of the most important fields of development which vitally affected all the others and radically changed the way of life of most people was that of transport. In the years 1750–1918 the improved road system, the establishment of a canal network followed by the railways transformed the country. Such transportation was the life blood of an efficient and successful industrialized society.

In the later eighteenth century, as the population increased and opportunities for work were more likely to be found at the factory bench than at home, the great migration of British people from country to town got under way. The process of urbanization, as it is now called, was a continuing one throughout the nineteenth and much of the twentieth centuries. It became a flood between about 1830 and 1880, a mass exodus which created an upheaval of such magnitude that at least 50 more years were to pass before society and the individual had adapted to it.

During the eighteenth century the new industrial factories were mostly still being built in rural districts where sites were cheap, raw materials were accessible and water power could

THE TELEPHONE.

7   *Advertisement for Professor Alexander Graham Bell's telephone of 1877 (page 156)*

8   *Edison speaking into his phonograph, 1899 (page 157)*

be utilized from the nearby fast-flowing streams. The transport system was also still largely dependent upon water – rivers, canals and the sea – so such sites were suitable for transportation of the factory's manufactured products. With the development of steam power and then the railways during the nineteenth century the picture changed. Concentration of industry in towns, many of which became important railway junctions, was found to be convenient and economical. Such towns grew at an unprecedented rate, causing untold problems in the too rapid expansion.

In 1700 London was Britain's only large city. During the following 200 years it grew very fast (the population quadrupling during the nineteenth century alone), attracting migrants from all over the country as industry after industry sprang up in the vicinity. But London, like the other ancient but smaller centres of population – York, Bristol, Norwich, for instance – was already an established city. These grew too fast for comfort but at least possessed some of the essential facilities needed in a town. It was in the villages and small towns which, almost overnight so it seemed, became cities, that many decades were to pass before a semblance of adequacy in amenities such as housing, water supply, sewage disposal, town administration and essential food and clothing supplies became available. Manchester, for instance, singled out by Friedrich Engels as a classic example of a city ripe for revolution*, had a population of about 8,000 in 1720; by 1900 it was over half a million. Birmingham's population rose from 70,000 in 1800 to half a million 100 years later. Typical of the more dramatic instances was Middlesborough, consisting in 1800 of a few farms, their inhabitants numbering 25; by 1900 its population was 90,000.

The old towns had grown up over many years around the cathedral or monastery, the castle and the trading centre of the market square. The industrial towns of the nineteenth century expanded round the factory districts. In the early years there was no city transport so the houses were built cheaply, closely stacked together so that the workers could walk to their place of employment. This meant overcrowding, dirt, inadequate light, air and sanitation and unimaginably bad conditions of living. The old rural cottages had been primitive and poor but, in the open space of the village, had been more bearable (pages 91 and 93). With this scale of rapid urban expansion, the established social fabric was destroyed: only slowly did a new one emerge. City populations became segregated, rich, bourgeois and poor districts developing. During the nineteenth century those who could afford it moved out further away from the city centre, to more spacious suburbs, leaving the poor breathing polluted air in hovels, clustered round the grimy factories.

By Edwardian times Britain had become an urban nation. Seventy per cent of the population lived in towns and the rate of increase had greatly diminished. Bad living conditions still existed for far too many people in the remote areas and the big cities but standards were gradually improving for many. England had led the world in urbanization, so creating acres of industrial slums. On the credit side, the country was also one of the world leaders in helping to eradicate these and housing the population in garden cities and new towns (page 100).

Historians hold differing views about the value of benefit and harm caused to society by the rise of technology and the resulting industrialization of the nineteenth century. How should the balance sheet read?

The belief that the Industrial Revolution was a catastrophe and that the introduction of factory methods of mechanization and rationalization exploited the workers, lowering their living standards, treating them as slaves and taking from them the dignity of craft labour, was widely held in the century between 1850 and 1950. The accounts of the horrors of the industrial towns, their factories, homes and poverty, written by Engels in the 1840s and Marx* in the 1860s were accepted as gospel truth by the socialist movement growing up in the later nineteenth century and these accounts served as a basis for many

The Condition of the Working Class in England by F. Engels. Published originally in 1845 in German (see bibliography).

Das Kapital by Karl Marx. Published 1867.

later writers. But Engels' description of working conditions in England is not entirely reliable. He wrote his book after only 20 months in the country and described many areas – Clydeside, the Potteries, South Wales, for instance – which he had hardly visited if at all. At best his book is an unbalanced account, contrasting, for example, the worst of Manchester's slums with the supposedly idyllic working conditions in English villages in the early eighteenth century.

More recent research since the Second World War lends more credence to Sir John Clapham's[*] contention, when he refers to the view that 'everything was getting worse for the working man' in the earlier nineteenth century, as a familiar legend. Certainly many intellectuals in England in the later nineteenth and early twentieth centuries appeared to hold too rosy a view of the life of a craftsman in pre-Industrial Revolution days. The master craftsmen and artists among these no doubt lived a satisfying life of independence doing interesting work, but this was the position of the very few. Most artisans carried out physically exhausting tasks for long hours with poor tools in bad conditions. The idea that under the cottage-industry system working conditions embodied freedom and dignity whereas only factory workers were exploited is a myth. In the textile industry, for example, the life of a cottage spinster or weaver in the early eighteenth century consisted of long hours of drudgery in poor light and chilly conditions to earn a pittance.

Technical innovation, mechanization and industrialization caused many evils and much suffering, but there is more evidence to show that these factors brought about an improvement in living standards than that they impoverished them. Professor Ashton cogently summarises the effect of the Industrial Revolution at the close of his book[†] when he says 'The central problem of the age was how to feed and clothe and employ generations of children outnumbering by far those of any earlier time'. He goes on to say that if England had remained a nation of cultivators and craftsmen she would have lost a considerable proportion of her population by emigration, starvation and disease.

MODERN LIFE

The social changes resulting from the technological developments of the eighteenth and nineteenth centuries were violent and complex. Few people escaped their effects and for two or three generations many people had to move their homes to an alien part of the country and adapt to a new environment and a totally different kind of work. In the twentieth century people's lives have been changed even more dramatically but in a less distressing manner. Since 1918 social policies have re-distributed the nation's wealth more evenly among the population and standards of living for most people have risen sharply. At the same time, technical achievement has transformed life in the home, office and factory, taken away the drudgery and need to work long hours, so making possible a fuller, more varied existence.

The pace of industrial and technical change is still accelerating. The scientific research of the late nineteenth century brought to fruition in the inter-war years many new developments: synthetic dyes, early plastics, electrical engineering, radio and television, travel by air, to name but a few. The trend towards a science-based technology has accelerated also, especially since 1945. The development of new industrial techniques is now undertaken to a large degree by large concerns and by government laboratories but, despite this, there is still place for the individual inventor. A study undertaken about mid-century showed that a considerable proportion of new inventions had come from such sources. The list included such varied items as the ball-point pen, the safety razor, chromium plating and air conditioning. New products initiated by the great industrial corporations, however, are often those which have required considerable capital investment. These include the development of float glass, nylon, the transistor, the integrated circuit (the silicon chip) and synthetic detergents.

We are now having to adapt our society to the challenge and problems of what is increasingly

[*] *An Economic History of Britain 1820–1929*, J.H. Clapham. University Press Cambridge 1950–2.

[†] *The Industrial Revolution 1760–1830*, by T.S. Ashton. Oxford University Press, 1966.

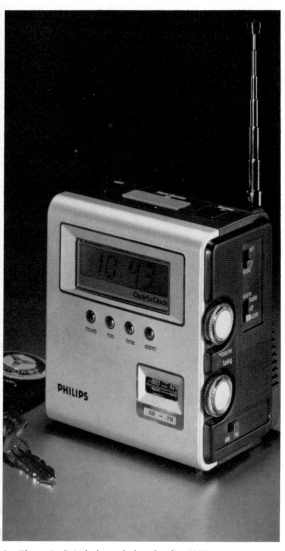

9   *Electronic digital alarm clock and radio, 1981*

being termed the 'second Industrial Revolution' (Chapter Nine). Electronics is making it possible to carry out a tremendous variety of mundane jobs by automatic control. Microchip miniaturization promises to bring possible benefits to almost every part of everyday life but its potential has yet to be realized. Such realization depends upon economic factors and it is by no means certain that people will appreciate and accept such 'benefits'. Already its effects can be noticed in improved television reception (satellite transmission), in facilities for telerecording, in more sophisticated music centres (page 160), in textile processes (page 148). Automatic scanning by probes in conjunction with a computer controls the books we take out from the public library, we collect our cash from an automatic device at the bank, laser scanning is planned to handle our supermarket purchases.

The full application of these new ideas would free us to live life to the full without spending time on routine work. The challenge is how we will spend our time. The problem is in employment for the community as a whole. By the year 2000 will people regard the second Industrial Revolution as having been a blessing or a curse?

CHAPTER TWO

Fuels and Sources of Power

The greatest single factor controlling the speed of technical advance has always been the source of power available at the time. The inventive genius of man has, at many stages in his history, produced useful mechanical and technological devices but frequently their effective development has been held up for lack of a suitable means of powering them. For instance, the mechanisms which were invented, among others in the early eighteenth century, to sow seed or to break up the soil were a great step forward from the previous hand-operated devices, but the most notable advance came only when the horse-drawn version could be replaced by the power of steam (page 27). Similarly, the hand-operated domestic washing machine or vacuum cleaner of the late nineteenth century represented remarkable household innovations, but their usefulness could not be exploited until the small electric motor had been perfected to power them (page 39).

It might be helpful, then, to give a preliminary chronological account of how and when such power sources became available with the fuels which made their use possible.

MAN AND ANIMAL POWER

The puny resource of the muscular power of one man was quickly supplemented in the ancient world by the concerted use of many men. There was no shortage of slaves in ancient Egypt to build the pyramids, or to man the mines and construct the roads of Imperial Rome. Slave labour, efficiently utilized, could provide a pleasant, cultured life for the privileged citizen.

In addition, a number of the basic aids towards lifting and moving heavy weights and carrying out tasks such as pumping water (10), crushing ore or grinding corn were well known to the ancient world. The principle of the lever was understood and its advantage utilized, also that of the spring, the wedge and the pulley. The cogged wheel, a system of gears and the screw were employed and developed.

Gradually man began to domesticate animals and harness their power. Donkeys and smaller animals had been worked since classical times, operating small vertical wheels on the treadmill or squirrel-cage principle to draw water from wells or grind corn. During the Middle Ages animals were employed in this way more widely in Europe for a variety of tasks (dog turnspits for example). The horizontal treadmill was then designed to utilize the energy of the heavier draught-animal, the horse and the ox, and this was more economical. Such animals powered all kinds of operations until they were superseded by the rotary steam engine in the early nineteenth century. Among other purposes they were employed to drive grinding and crushing machinery, to pump out mines, to raise coal, to make butter and operate textile machinery (11, 12). The treadmill worked on the wheel and axle principle; as time passed, a number of different designs were produced to improve efficiency.

10   *Treadwheel water pump, 1556. Agricola*

11   *Great Exhibition Catalogue, 1851*

BARRETT, EXALL AND ANDREWES'
PATENT TWO HORSE THRASHING MACHINE WITH PATENT GEAR WORK,
AS IN OPERATION.

12   *Horse-powered tanner's mill for grinding bark. From W.H.
Pyne's 'Microcosm', 1803*

WATER POWER

The use of a water-wheel to tap the energy of flowing water dates from antiquity. The earliest type (called a Greek or Norse water-mill) consisted of a vertical wooden axle to the lower end of which were fitted wooden vanes; these were immersed in and turned by the stream. Its chief use was for grinding corn.

The more familiar water-wheel derives from Vitruvius' design of the first century BC. In this type the axle is horizontal and the wheel vertical. Such large wooden wheels had an outer and inner rim with blades fitted between them. The earlier designs were of the undershot type, that is, the lower part of the wheel was immersed in the stream so that the force of water turned the wheel (13). Water-wheels were an important source of power in the Roman Empire. They were used, for example, for flour milling, irrigation and pumping out mines, producing far more power than animals or men.

The water-wheel returned to Europe in the early Middle Ages after several centuries when its use had been forgotten. By 1600 it had become the most important source of motive power, employed for all kinds of industrial purposes from mining, textile processes, timber and metal mills to tanning and irrigation. It continued in use throughout the nineteenth century, though iron gradually replaced wood as the material for its construction. Indeed, the terms steel mill and cotton mill, for example, resulted from the years when industry was powered by water.

An alternative design was introduced during the Middle Ages; this was the overshot type in which the water fell on to the upper part of the wheel. The blades or paddles were constructed into triangular troughs or buckets which filled with water. The weight of this water caused the wheel to turn, against the stream, in the opposite direction from the movement of the undershot design. This overshot wheel was found to be more efficient but it did require more complex installation as the water had to be diverted and directed on to the upper half of the wheel (14). In the eighteenth century John Smeaton (1724–92), the civil engineer (page 98), carried out extensive experimentation into the relative merits of the different types of wheel. He found that, with two wheels of identical diameter, over 60 per cent of the possible energy supply from the flow of water was utilised in an overshot design but only 22 per cent in an undershot one.

WIND POWER

The origin of the windmill is obscure. It is often presumed that its action stemmed from the knowledge gained in sailing ships: a reasonable but unproven theory. This action comprises the combination of sails powered by the wind which then rotate to turn a treadmill to grind (mill) the flour. The windmill evolved at a later date than the water-wheel but, similarly, there were two basic designs: one structured on a vertical axis, the other on a horizontal one. Again, as with the water-wheel, the earlier, vertical-axis pattern originated in the Near East (this time in Persia some time before the tenth century), while the later, horizontal-axis type (possibly inspired by Vitruvius' water-wheel) came from western Europe, where it was first mentioned about 1185 and was in common use in the following century.

Windmills were widely employed for grinding corn and later for raising water until well into the nineteenth century, though their use was more limited than that of water-wheels. They were to be found chiefly on the plains of northern Europe and, in Britain, in the flatter, open, windy areas on the eastern side where rivers tended to flow too sluggishly to be suitable for water power. The western design proved more efficient and, as time passed, became more elaborate as various devices were perfected for turning the sails into the wind; the mechanical fantail of 1745, for instance, which automatically adjusted the direction of the mill as the wind changed.

The earlier models were post mills. In these the whole structure had to be turned into the wind by means of a tail pole attached at the rear. The miller needed considerable strength to shoulder the whole windmill round by this means. The box-shaped wooden body carried sails on a nearly horizontal shaft and this was supported and pivoted on a massive, well-anchored, central vertical post (15).

The post mill was followed in the late fourteenth century by the tower mill. This had a cap

13   *Corn mill, undershot water-wheel, 1662*

14   *Corn mill, overshot water-wheel, 1662*

15   *Corn grinding post windmill, 1588, Ramelli*

16   *Tower windmill for raising water*

which contained the shaft on which the sails revolved and it was only this cap which needed to be turned. The stationary body of the windmill could then be made of brick or stone which was more durable and provided better wind resistance (16).

CHARCOAL-BURNING

This was a craft of ancient origin which provided a useful fuel until the widescale development of the coal and coke industries in the late eighteenth century. Charcoal gives about twice the heat of an equivalent weight of wood so it is less cumbersome to handle. It was used for centuries as a domestic fuel both for heating and cooking. For the Romans, for example, it was the preferred cooking fuel. Charcoal was also widely employed as a fuel in the smelting of iron (page 67), partly for the high temperatures which it would give but also because of its low sulphur content.

The manner of charcoal-burning altered little over the centuries. The most usual method was to build up a structure with cut lengths of wood which formed a central flue. Further lengths of wood were then piled around the sides making a cone. The outer surface of this cone-shaped kiln was then covered with a layer of mixed earth and powdered charcoal to exclude the air and so hold the combustion down to a minimum level as too fast burning would reduce the wood to ash. The kiln was then lit by dropping glowing embers of charcoal down the flue, after which the hole was sealed with turf and earth.

The kiln would take several days to burn the wood to make charcoal. All this time, night and day, it had to be watched in order to control its speed of burning. For this reason charcoal burners and their families lived in huts on the site in the woods, away from centres of habitation. Such a camp has been set up at the Weald and Downland Museum at Singleton in Sussex by two retired charcoal burners from Horsham. For hundreds of years the Sussex Weald was an important charcoal-burning area, providing the fuel for the iron industry there.

COAL-MINING

It was Britain's abundance of coal reserves which made her early industrialization possible (page 15); yet the mining of coal only dated back a few hundred years. Metals such as copper, lead and gold had been mined in antiquity but the importance of coal as a fuel and as a constituent in western technology is comparatively recent. The rich outcrops of coal seams in Britain were known to the Romans, who worked these surface deposits as a fuel for heating, but not extensively. After the departure of the Romans there is virtually no mention of the use of coal until the twelfth to thirteenth centuries. Even then it was only readily available in colliery districts and in coastal towns accessible to the sea-coal supplies brought in by small boats plying from the collieries. Transportation was costly and mining difficult.

It was the depletion of forest supplies (with little replanting) for fuel, ship-building and manufacturing industry, together with an expansion of population, which led to restrictions upon the burning of wood in the later sixteenth and early seventeenth centuries. From about 1600 until the later nineteenth century coal was the chief fuel for all purposes: manufacturing, military, domestic. This substitution of coal for wood led to important advances in the technology of burning the fuel, getting the best use from the resource for whatever the product: glass, bricks, salt, etc. New furnaces and new manufacturing techniques had to be devised. Deeper mining presented the greatest problems, not least in pumping out the water to avoid flooding, and this need led to the eighteenth-century development of the steam engine (page 28).

By the end of the eighteenth century coal-mining had become an established industry vital to the expansion of manufacturing and to the development of industrial technology which was creating the wealth to make Britain a leading nation. Coal was supplying the fuel to heat water to provide the power to work the new machines. The mechanical inventions and the increased output of coal reacted upon one another, each new invention leading to further expansion.

The key to this sustained advance was a tremendous increase in coal production. Between 1750 and 1850 the increase was ten-fold, by 1914 it had further multiplied four times. The prob-

17 *Moving coal from the face by female and child labour. Early nineteenth-century mine*

lems created by this speed of advance were manifold and decades were to pass before they were solved. Steam power had eased the problem of flooding, but it was much more difficult to use such powered machinery at the coal-face.

Until the last quarter of the nineteenth century increase in output continued to depend upon the use of ready supplies of cheap labour, human and animal. Coal was cut by hand-pick and crowbar. It was blasted by gunpowder but no safety fuse was available until 1831. In the eighteenth century women still carried coal in baskets up ladders from gallery to gallery and as mines became larger trucks were man-hauled along rails. Pit ponies worked underground from 1763. The hazards of gas and explosion were immense.

Only slowly were conditions improved. The Davy miner's lamp came in 1816, steam haulage was introduced from the 1820s and finally coal-cutting machinery was developed. The coal-cutter of 1863 introduced the principle of rotary cutting machinery which led the way towards a number of more advanced models, greatly increasing the cutting capacity, though it was after 1900 before mechanization at the coal-face made notable progress. Increased capacity in cutting made urgently necessary a mechanical means of collecting the coal and carrying it away; the face-conveyor of 1902 was an early answer to this. In the 1930s and 1950s much more advanced equipment followed: coal-cutters, conveyors, hydraulic roof props, centrifugal pumps as well as much improved ventilation, lighting and safety methods. The ultimate aim, now much nearer, was totally automated mechanized mining.

POWER FROM STEAM

It was in 1769 that James Watt, whose name is familiar to every schoolchild in connection with the making of a workable steam engine, patented his separate condenser, the invention of which led to the wide-scale use of steam as the prime mover of the Industrial Revolution. Since before 1600 mining engineers had desperately sought such a power source to pump out flooded mines and provide adequate ventilation. The existing sources – animals, wind, water – were proving inadequate as mines became larger and deeper. Many attempts were made to develop the power that it was known could be generated by a jet of steam.

The study of the behaviour of gases, which includes air, had interested scientists since classical times. The use of bellows to produce a current of air was known in Hellenistic Greece and adapted later for mine ventilation and to increase the heat from fires used in iron smelting; the bellows were powered by horses or water-wheels (18). How to make use of compressed air to produce sound from organ pipes was another Hellenistic discovery, again adapted later in many practical ways. Hero, the Alexandrian physicist and engineer, produced his aeolipile in the first century AD: an instrument intended to be operated by steam.

Over the centuries a variety of means was found to raise water for irrigation and drainage, ranging from the Archimedean screw of antiquity to the medieval suction pump which operated by utilizing the weight of the air in the atmosphere. Such pumps were initiated in fifteenth-century Italy and came to be widely used, but even by the seventeenth century, though mining engineers knew that it was not possible to draw water from a depth greater than about 30 feet, they did not understand why. In 1641 the engineers of Cosimo de' Medici appealed to Galileo to help them design a pump which would draw up water from 50 feet below. Galileo believed, as scientists had done since the days of Aristotle, that a vacuum could not exist in nature (indeed, this is still so on the surface of the earth without the intervention of man). That 'nature abhors a vacuum' was currently understood to be the reason behind the workings of a suction pump. Galileo had been among the first to experiment with a vacuum but, at that time, could only conclude that 'there was a limit to nature's abhorrence'. It was his pupil, Evangelista Torricelli (1608–47), who discovered that the atmosphere exerted a pressure because of its weight and that this corresponded to a column of water just over 30 feet in height.

Torricelli's discovery aroused great interest in other scientists who then experimented with the use of vacuum and the power of the atmosphere. Among these classic experiments were the colourful demonstrations of Otto von Guericke (1602–86) (20), the German scientist who was the Mayor of Magdeburg. He designed several vacuum pumps and in 1657, using one of these, evacuated air from a copper sphere, some 20 inches in diameter, made from two halves which had been sealed together. He demonstrated the

immense force due to the atmospheric pressure by showing that two teams of eight horses could not pull the hemispheres apart (3). In another experiment, more nearly relevant to the action of the steam engine, he used a piston closely fitted into a 15-inch diameter cylinder. He showed that if most of the air was removed (i.e. a partial vacuum was produced) below the cylinder, a group of 50 men hauling on ropes were not able to offset the piston being driven into the cylinder by the pressure of the atmosphere (19).

It was Denis Papin (1647–1712), the French Huguenot scientist, who carried the investigation an important stage further (21). From his many researches and experiments came the first pressure cooker (the digester of 1679), the double air pump and the air gun. In 1690 he pursued his idea of condensing steam to make a vacuum by making a working model which established the principle that steam could be used to move a piston in a cylinder. It was this principle which was then made to work in the early steam engines which followed. Papin's experiment consisted of a small cylinder containing a piston. He boiled a little water at the bottom, the steam from which raised the piston to the top, where it was retained by a catch. The source of heat was removed and the steam condensed, producing a partial vacuum. The catch was then released and the piston was driven down into the cylinder because of the pressure of the atmosphere.

The early practical working steam engines were really atmospheric engines since atmospheric pressure was their main power source. The first of these was designed by Thomas Savery (c. 1650–1715), a military engineer and inventor from Devon, who intended his equipment to pump out water from the tin and lead mines of the West Country. He demonstrated to the King in 1698 a working model of his 'Engine to raise Water by Force of Fire' and took out a patent for it. The principle of this pump was to raise water to as high as the atmospheric pressure would allow by the condensation of steam in closed vessels (to produce a vacuum) then, by the pressure due to the super-heated steam, to raise the water higher still. The technology of the time was not good enough to manufacture vessels which could withstand this steam pressure of several atmospheres and Savery's pumps were

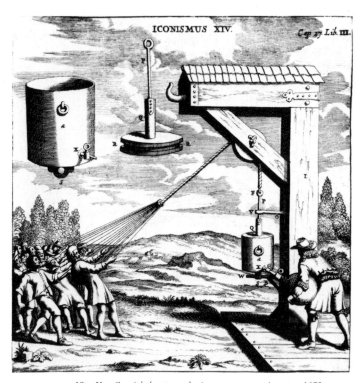

19   *Von Guericke's atmospheric pressure experiment, c.1672*

18   *Bellows for ventilating mines, powered by horses, 1556.*
Agricola

20   *Otto von Guericke*

21   *Denis Papin, 1711*

not reliable for the purpose he intended. They were used, however, for many years to raise water for supply in large houses and to water-wheels, also to turn machinery.

It was Thomas Newcomen (1663–1729), an obscure ironmonger from Dartmouth, who produced, after 15 years of experimental work, an atmospheric engine whose line of development stemmed directly from the work of von Guericke and Papin. In his engine of 1712, constructed to work a water-pump in a colliery near Dudley Castle, he took a step further Papin's idea of a piston driven upwards in a cylinder by the force of steam and downwards again by atmospheric pressure. His cylinder was supplied with steam from a boiler beneath. The sealed piston was attached by a chain to one end of a great beam above, the other end being connected to the coal-mine water-pump below. As the steam pushed the piston upwards the water pump in the mine was made to work. The steam valve below the cylinder was then closed, a spray of cold water was injected into the cylinder, the steam condensed and the vacuum created caused the atmosphere to force the piston downwards and the mine water-pump continued to operate. The piston was raised again by the overbalancing weight of the pump rod. Newcomen's engine was not very efficient (it used a great deal of coal and developed about six horse-power) but it was the first practical, reliable design and so a great success. It was widely used in England and introduced on the Continent and in America (23).

Chief credit for transforming the steam engine into the prime mover for nineteenth-century industrial needs must go to James Watt (1736–1829) (25). The idea of the separate condenser was due to the inventiveness and scientific curiosity of this young mathematical instrument maker at Glasgow University. Encouraged and inspired by his friend Professor Joseph Black there, who laid down the scientific principles, and aided by the enterprise and business acumen of his partner (from 1775) Matthew Boulton, owner of the Soho Engineering Works near Birmingham, and by the engineering skills of the great ironmaster John Wilkinson, Watt spent most of the rest of his life perfecting his steam engine, incorporating one innovation after another to provide the power which was eventually to transform industry.

Watt's interest was first aroused when he was asked to repair a model of an atmospheric engine. He noted its inefficiency and waste of steam. Concluding that this was due to the necessary heating (to 212°F) and cooling (to 100°F) of the cylinder between each stroke, in 1765 he worked out an idea for keeping the cylinder containing steam separate from the condensing vessel. His patent of 1769 included this separate condenser. Watt went on to make several models, but it was in 1776 that Boulton and Watt produced the first two full-size engines. In these the cylinder was insulated to keep it at a temperature of 212°F and the condensing vessel was fitted with an air pump to maintain the vacuum by pumping out condensed water and air from it.

Boulton and Watt continued to make these single-action, reciprocating beam engines for some years, but their action was only suited to pumping up and down. Boulton was alive to the urgent need of heavy industry for a rotative steam engine which could be used to power all kinds of machinery. He knew that he could sell such engines. Watt worked on a number of ideas for converting his reciprocating engine into a rotative one and in 1781 patented his sun-and-planet idea (so-called because the planet-wheel moved round the sun-wheel). This engine was soon being built in numbers to turn heavy machinery in, for example, iron foundries and flour and textile mills.

Soon followed Watts' double-acting rotative engine in which the force of steam moved the piston first in one direction then the other as it was admitted alternately to each end of the cylinder; this was an immense advance in the efficient use of fuel and steam. In 1788 he introduced his centrifugal flyball governor, linking it to the steam inlet valve to control the speed of his engine. This device had been in use for some time in windmills, controlling the adjustment of grinding stones according to wind velocity but Watt was the first to apply its use to the steam engine in a feed-back capacity* (22, 24).

James Watt retired in 1800 on the expiry of his

---

* Feed-back is a twentieth-century term, its action not mathematically analysed until 1927. It is a self-regulating control system widely used in modern equipment (in which a fraction of the output is fed back into the input.

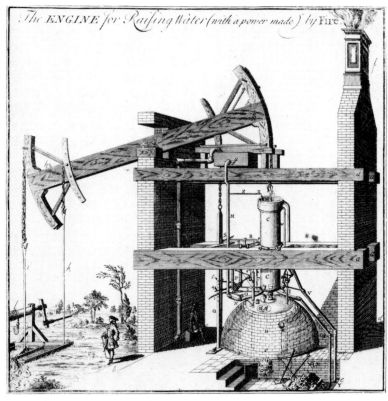

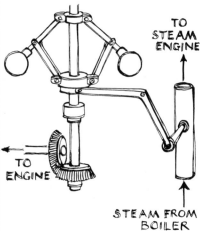

22 *Flyball centrifugal governor of the type used by Watt*

23 *Beighton's engraving of a Newcomen engine, 1717. B boiler, C cylinder, I pump rod to mine*

24 *James Watt's rotative engine, 1788*

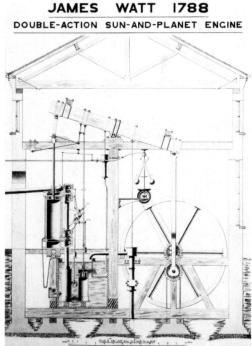

25 *James Watt. Engraving after painting by Beechey*

patents. The firm of Boulton and Watt had made 496 Watt engines, 60 per cent of which were rotative. Steam engines were at work all over Europe and America and as this power became more widely employed a need arose for a standard for engine capacity. Watt suggested horse-power as a term, the working capacity of a horse being familiar to everyone. As a result of his studies of working draught horses in London he decided that one horse-power was equivalent to lifting 550 lb weight one foot per second. Much later, with the development of electricity as a power source (page 39), James Watt was remembered when a watt became a unit of electrical power (one horse-power = 746 watts).

It was in the nineteenth century that steam power came into its own. When Watt had begun to build steam engines in the 1770s he was handicapped by the lack of skilled craftsmen to make such equipment with sufficient accuracy. They had to be built in wood and iron by local wheelwrights, carpenters and blacksmiths. Only gradually were specialist engineers trained to build the engines.

The next advance towards higher thermal efficiency was by the introduction of high-pressure steam which Watt had not dared to apply because of the materials and quality of engineering available to him. After the expiry of Watt's patent, Richard Trevithick (1771–1833), a Cornish mine engineer, had built a small pumping engine in 1802 in Coalbrookdale. Only four feet in diameter, its boiler was one and a half inches thick and developed a steam pressure ten times that of atmospheric pressure. Trevithick built many high-pressure steam engines, stationary and locomotives. There were problems, but he was soon followed by others and high pressure steam power revolutionized the industrial scene in manufacturing industry and transport (6).

In the second half of the century, late in the day, came scientific study into heat; the steam engine is a device for converting heat energy into mechanical energy and so governed by the laws of thermodynamics which were enunciated as a result of the work of such physicists as J.P. Joule (1818–89), R.J.E. Clausius (1822–88) and Lord Kelvin (1824–1907).

With greater understanding, steam engines became more powerful and efficient. The steam turbine was developed towards the end of the century and in this, steam, despite gradual replacement by other power sources, still plays a vital part in power station generation, whether coal- or oil-fired, or employing nuclear fuel (page 42, and 43, 44).

GAS

Not until the late eighteenth century was gas seriously considered as an illuminant. Professor Minckelers (1748–1824) experimented with lighting his lecture room at the University of Louvain in 1784 with gas derived from coal. In 1801 Philippe Lebon (1767–1804) gave a public exhibition in Paris of the possible use for lighting and heating by gas made from wood. It was the work of William Murdock (1754–1839), a Scottish mechanic employed by the engineering firm Boulton and Watt, whose research into coal gas led the way to its commercial application. Murdock lighted his home in 1792 with gas produced in iron retorts and then conveyed through metal pipes. He went on in 1798 to install a lighting system at the firm's Soho Works. By 1804 his apparatus was sufficiently advanced for the firm to canvass orders.

Murdock's experimentation was followed by others. Samuel Clegg (1781–1861) attempted a purification of gas and installed a lighting system in some factories. Frederick Winsor (1763–1830) foresaw the possibilities of piping gas long distances to light the streets. His company, later the Gas, Light and Coke Company, demonstrated the first public installation of gas street lighting in London in 1807 by illuminating part of Pall Mall. By the 1820s miles of London streets were lit and gas illumination was being installed in shops, banks, churches, theatres, clubs and public buildings, but it was the second half of the century before gas was satisfactorily adapted for cooking and heating (pages 117 and 120).

Gas was a predominant fuel for illumination, cooking and heating for many years in the nineteenth and twentieth centuries and during this time it was mainly derived from coal. The discovery in 1965 of natural gas under the North Sea gave a boost to the industry, extending its use as a fuel tremendously in the succeeding decade.

PETROLEUM
Although surface petroleum deposits had been known since antiquity in Egypt and Mesopotamia, as well as further east, their use over the centuries was limited to medicinal needs and the making of bitumen, paints and varnishes. The word 'naphtha', coined in Persia from a Greek term, described a material which flamed and was experimented with as a lamp fuel.

In Europe it was not until the late eighteenth century that, as in the case of gas, interest was seriously aroused in the possibilities of using petroleum for illumination. As a result of the industrial revolution and its consequent establishment of factory working and urbanization (page 17) a better quality of lighting and a means of providing artificial illumination during the hours of darkness, especially in winter, presented a problem requiring an urgent solution.

It was in the years 1780–1840, when gas lighting was in its infancy or available only in urban areas, that an alternative form of illumination was badly required; the animal and vegetable oils so far in use were proving inadequate for the purpose. Various petroleum products were discovered and developed. James Young (1811–83), a scientist who had assisted Faraday at the Royal Institution, began manufacture of paraffin oil from shales in Scotland. Meanwhile Abraham Gesner (1797–1864) perfected a means of producing kerosene (the American term) from petroleum. He named this substance from the Greek word *keros* = wax, but in England it is called paraffin.

This new oil kerosene sold well in America in 1856, but this sale was soon overtaken by events. A year later oil wells were being drilled in Hanover in Germany and dug in Ploiesti in Rumania. In 1859 came important successful drillings in Pennsylvania in the USA, which proved a historic turning point in petroleum production. At this time and, indeed, for the remainder of the nineteenth century, the requirements were for illumination, heating (page 120) and lubrication; petrol was a dangerous by-product which only became of value with the invention of the internal combustion engine.

In modern times petroleum has become not only a vital source of energy needed for transport, heating and a general power source but its products and by-products supply multiple needs in the civilized world from lotions, laxatives, polishes, sprays and ointments to a complex range of plastics and detergents (pages 79 and 83, also 26).

ELECTRICITY
Electricity is now a widely-used form of energy, yet for centuries man was unaware of its existence. Only slowly did understanding come of its nature and potential, while it has been little more than 100 years since techniques began to be developed to harness its power, so transforming life in industrialized countries. In modern times, of all forms of energy, it is the most pervasive, being at the heart of communication, lighting and heating systems: indeed, it is the basis of the supply of all our everyday needs.

The phenomena of magnetic attraction, as evidenced in the lodestone (magnetic oxide of iron) and that of electric attraction, when a substance such as amber was rubbed with wool, were observed by the ancient Greeks. The English word magnet derives from the Greek name for the black lodestones found in Magnesia in Asia Minor, while electron is the Greek word for amber. Little further understanding was achieved until the European Renaissance when navigational needs aroused interest. In England, William Gilbert (1544–1603) published the results of his years of study of electricity and magnetism in his *De Magnete* of 1600. Sixty years later von Guericke's experiments led to his invention of a frictional machine which produced sparks. In 1709 Francis Hawksbee produced a faint glow of electric light by rotating an evacuated glass vessel and simultaneously rubbing it by hand.

With the eighteenth century came a number of interesting discoveries in Europe and America which added to the sum of knowledge and brought the future generation of electricity imperceptibly nearer. For instance, Stephen Gray (c.1666–1736) in England elucidated the principle of the conduction of electricity and established that some materials were conductors

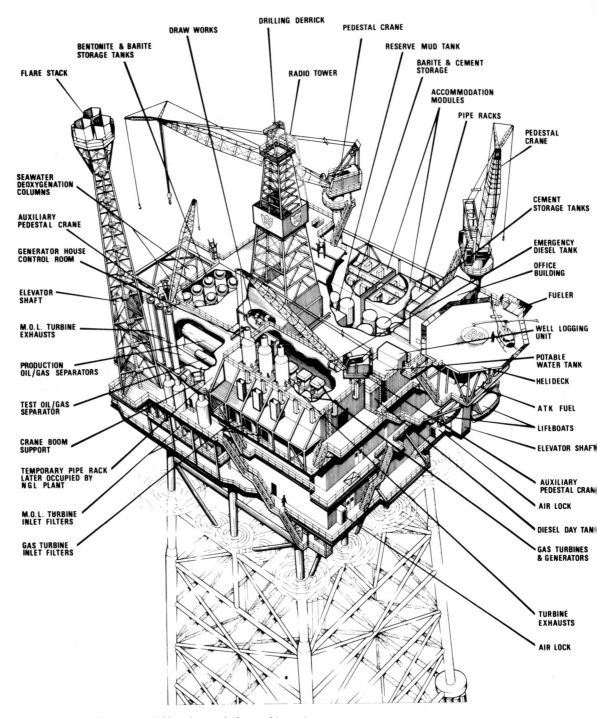

FLARE STACK

BENTONITE & BARITE
STORAGE TANKS

DRAW WORKS

DRILLING DERRICK

PEDESTAL CRANE

RESERVE MUD TANK

RADIO TOWER

BARITE & CEMENT
STORAGE

ACCOMMODATION
MODULES

PIPE RACKS

PEDESTAL
CRANE

SEAWATER
DEOXYGENATION
COLUMNS

AUXILIARY
PEDESTAL CRANE

GENERATOR HOUSE
CONTROL ROOM

ELEVATOR
SHAFT

M.O.L. TURBINE
EXHAUSTS

PRODUCTION
OIL/GAS SEPARATORS

TEST OIL/GAS
SEPARATOR

CRANE BOOM
SUPPORT

TEMPORARY PIPE RACK
LATER OCCUPIED BY
NGL PLANT

M.O.L. TURBINE
INLET FILTERS

GAS TURBINE
INLET FILTERS

CEMENT
STORAGE TANKS

EMERGENCY
DIESEL TANK

OFFICE
BUILDING

FUELER

WELL LOGGING
UNIT

POTABLE
WATER TANK

HELIDECK

ATK FUEL

LIFEBOATS

ELEVATOR SHAFT

AUXILIARY
PEDESTAL CRANE

AIR LOCK

DIESEL DAY TANK

GAS TURBINES
& GENERATORS

TURBINE
EXHAUSTS

AIR LOCK

26   *North Sea Oil. BP Forties Field production platform and its equipment*

and some insulators. In 1745 the Dutch professor of physics at Leyden discovered how to collect and store an electric charge which had been produced by friction; the device known as the Leyden Jar. At the end of the century occurred the important Italian contribution when Count Alessandro Volta (1745–1827) correctly interpreted the findings of Luigi Galvani's experiments on twitching frogs' legs and produced the first chemical battery (27). This was his 'voltaic pile' of 1800, consisting of a stack of alternating zinc and silver plates separated by cardboard sheets soaked in brine. It was a great step forward as it provided a simple means of producing a continuous electric current for experimental purposes.

The publication of Volta's work stimulated interest in electrical research by scientists in many countries. The Danish physicist Hans Christian Oersted (1777–1851), searching for a connection between electricity and magnetism, published in 1820 his observations of the deflection of a magnetic needle by an electric current passing through a nearby conductor. Five years later the Englishman William Sturgeon (1783–1850) built the first practical electromagnet. The French physicist, André Marie Ampère (1775–1836), meanwhile was experimenting on electromagnetic forces between conductors carrying electric currents (28) and the German, Georg Simon Ohm (1789–1854), published in 1827 his famous Law which established the relationship between voltage and current strength and led to the idea of electrical resistance.

The decisive step forward which proved to be the turning point in the understanding of electrical theory and led directly, though much later, to the manufacture of generators and electric motors was Michael Faraday's (1791–1867) successful induction of electric current in 1831: an experiment which he described to the Royal Society in London on 24 November of that year (30).

Faraday, who was familiar with Oersted's demonstration showing that magnetism could be obtained from electricity, wondered whether the converse might be true and magnetism could produce electricity. As early as 1822 he wrote in his notebook: 'convert magnetism into electricity'. Between 1825 and 1831 in his laboratory at the Royal Institution, he made several attempts to do this and eventually, on 29 August, 1831, by winding two coils of wire round opposite sides of an iron ring (insulated from each other and the ring) sent a current into the first coil from a battery but no current reached the second coil. Then he discovered that at the moment when the battery was connected or disconnected a momentary current was created (29). He experimented further and found that intermittent surges of current could also be obtained by various forms of motion of a magnet or conductor relative to one another, as, for example, when he thrust a bar magnet into his coil of wire and, again, when he withdrew it, or if he moved the coils in relation to a magnet (29). Finally, on 28 October, to produce a continuous current, Faraday set up an apparatus comprising a copper disc rotating between the poles of a permanent magnet: here was the first electric generator (29).

Faraday's devices were experimental. He went on to enunciate the laws governing his discoveries and to show that electricity derived from friction and that from his generator, were the same. He then turned to research.

Faraday had made the essential discoveries, but there was no immediate industrial need for these. Other scientists continued experimentation and improved batteries were able to supply the then electrical requirements. It was the 1870s before practical efficient generators were made and a decade later for electric motors. The history of the development of electricity is the reverse of that of steam. In the former, scientific research came first, followed much later by technical achievement; in the latter, empirical experiment long antedated scientific theory.

Only a year after Faraday's demonstration, the first magneto-electric machine constructed on his principles was shown in Paris by Hippolyte Pixii. It was followed, over the years, by contributions from scientists and engineers in many countries but, not until the 1870s did the Belgian inventor Zénobe Théophite Gramme (1826–1901) construct the first practical efficient dynamo to produce a satisfactorily steady current. It was chiefly designed to be driven by steam engines

27 *Count Alessandro Volta*

28 *André Marie Ampère*

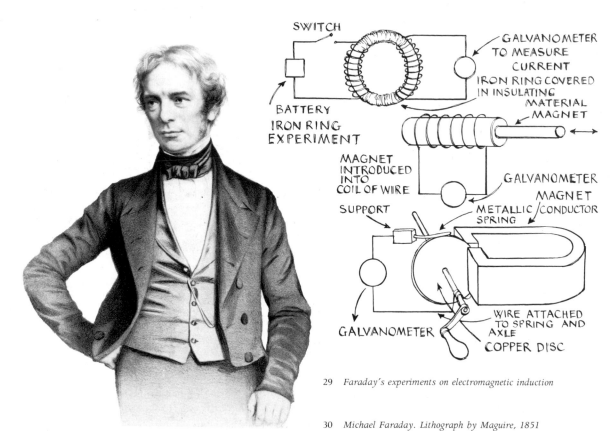

29 *Faraday's experiments on electromagnetic induction*

30 *Michael Faraday. Lithograph by Maguire, 1851*

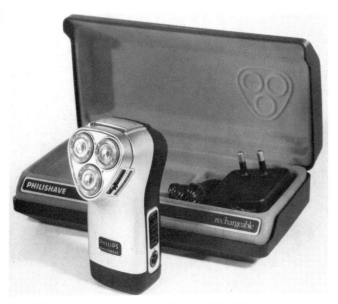

31 *Philips Philishave electric razor, 1981 (type HP 1312)*

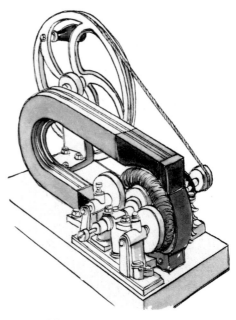

32 *Hand-driven Gramme generator c.1870*

33 *Bürgin dynamo of type made by Crompton and Co., 1881–2*

34 *Nikola Tesla*

35   *Fan powered by electric motor. Merryweather's Catalogue c.1900–10*

36   *Universal electric mixer-beater, 1918. Double rotary mixer with attachements*

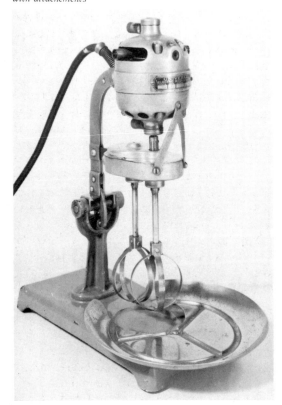

37   *Thor washing machine and wringer made c.1920*

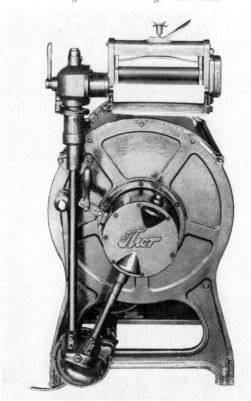

producing continuous current on demand. Improvements were later made by the Swiss engineer Emile Bürgin, and R.E.B. Crompton of England (32, 33).

By the middle of the nineteenth century a range of mechanical devices was being designed in an attempt to relieve domestic drudgery on the preparation of food, the cleaning of the home and the laundering of clothes and linen but all of these had to be hand (or foot) operated. Such 'labour-saving' machines were generally of complex design comprising wheels, cogs, treadles and levers to beat, peel, core, stone or slice all kinds of food, suck or sweep dirt from carpets or wash clothes in a tub which still had to be filled and emptied by hand. The introduction of an electric motor suitable to power such devices was, arguably, the greatest single event in the history of housewifery; during the twentieth century it has revolutionized every domestic task. The principles governing the electric motor, similar to those of the dynamo, had been shown by Faraday. Large motors were available from the early 1880s, but what was needed in the home was a smaller, more convenient design. In 1889 Nikola Tesla (1856–1943), the Yugoslav-born American inventor of the prototype AC motor (34), designed the first small one-sixth h.p. motor for the American company Westinghouse who put it on sale to drive a domestic fan (35). Later, such power units were adapted for vacuum cleaners and washing machines, but it was not for some time that suitably small motors were fitted into portable household equipment. For example, one of the very early vacuum cleaners powered by electricity was the Trolley Vac designed by Booth (40) and produced in 1906 by his Vacuum Cleaner Company (38) but this elaborate, expensive equipment (35 guineas), contained on a tray on a trolley which was pushed around the house and comprised an electric motor which drove a rotary vacuum pump by means of a belt, weighed nearly a hundredweight so was impracticable to take upstairs. It was 1910 before motors were fitted to make such cleaners a commercial success.

Since the 1920s a household mechanization based upon the power of the electric motor has been taken increasingly for granted (36, 37, 39, 41). Motors have been steadily miniaturized so that by the 1960s they were fitted into the smallest of kitchen apparatus to power all the devices upon which we now depend from lawnmowers and 'fridge-freezers' to shavers and food processors. More recently, electronic control of temperatures and computer programming has become more important in the design and function of this equipment (Chapter Nine).

Over the years, units and measuring devices in electricity, many of which are household words, have been named to honour those scientists who have particularly contributed to the knowledge and advances. The amp (Ampère) is a unit of electrical current, the farad (Faraday) a unit of capacitance, a galvanometer (Galvani) measures small amounts of electricity, an oersted (Oersted) is a unit of magnetic field strength, an ohm (Ohm) a unit of resistance, a tesla (Tesla) a unit of magnetic flux density, a volt (Volta) a unit of electro-motive force and a watt (Watt) a unit of power.

NUCLEAR POWER

This source of energy is different from those discussed previously: it is based on the fission of the atomic nucleus or the fusion of nuclei, whereas the older sources derive from the molecule (which is at least 100,000 times bigger than the nucleus). In both fission and fusion the mass (masses) of the nuclei before the process is greater than that after it has taken place. The mass loss becomes energy. Unlike, for instance, the burning of coal, the process uses up only a very tiny quantity of the naturally-occurring material to produce a vast amount of energy: 1 lb of uranium (which is about 1 cubic inch) is able to release as much energy as the burning of 1,000 tons of coal.

All the forms of energy previously utilized by man derive from the sun, which creates the wind to drive the windmill, the motion of the water for the water-wheel, the coal, oil and gas which all come originally from plants and animal life. When men made steam engines or electric motors to convert one form of energy into another (e.g. electrical into mechanical) the atoms were not altered; now in the production of nuclear energy the atom itself is substantially altered.

The word 'atom' comes from the Greek *atomos*

38   *Maid using Booth's Trolley-Vac, 1906. Note: clean patch of carpet just vacuumed and apparatus connected to electric light fitting*

which means a particle of matter so small that it is not conceivably further divisible. Democritus, the Greek philosopher of the fifth century BC, used the word when he described his belief that the world consisted of an infinite number of tiny, indivisible yet eternal particles.

From the time of Dalton's atomic theory (page 16), through the Italian Count Avogadro's study of molecules (1811) to Wilhelm Röntgen's (1845–1923) discovery of X-rays in 1895 which led to J.J. Thomson's (1856–1940) identification of the electron in 1897, to Pierre and Marie Curie's discovery of radium in 1898 and to the New Zealand-born Lord Rutherford's (1871–1937) experiments of the early twentieth century, a much clearer picture began to emerge of what the atom was like. By 1911 Rutherford had evolved his theory of the structure of the atom.

This picture showed the atom no longer as a tiny piece of solid material but a complex arrangement which consists of a central body (named by Rutherford the nucleus), revolving around which are negatively-charged particles (electrons): a kind of miniature solar system. The nucleus, which is an atomic nerve-centre, is made up of different particles, non-charged ones called neutrons[*] and ones with a positive electric charge (protons). Atoms of all elements contain the same type of particle structure, but what varies is the number of protons and electrons[†]. Hydrogen is the smallest and simplest atom, containing one

[*] This picture of the atom was made more complete by the discovery in 1932 of the neutron by Sir James Chadwick (1891–1974).

[†] Normally each atom contains an equal number of protons and electrons so the charges cancel each other out leaving the atom electrically neutral.

39   *Colston Dishwasher, 1976*

40   *(Bottom right) Hubert Cecil Booth, construction engineer and founder of the Vacuum Cleaner Co. Ltd (now Goblin BVC Ltd). Mr Booth coined the term 'vacuum cleaner' for his first suction device patented in 1901*

41   *Hoover carpet shampooer/floor polisher and scrubber. Plastic and metal, 1979*

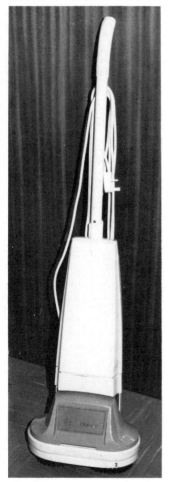

electron and one proton. In nature uranium is the most complex and contains 92 of each.

It is now known that the structure of the atom is more complex than the simple model defined by Rutherford, but it was from this basis that nuclear power developed. It was understood before the First World War how great was the potential energy contained in the atomic nucleus: the problem was how to release it. Scientists in many countries worked on the question during the 1920s and 1930s, notably Rutherford and Cockcroft in England, the Joliot-Curies in France, Neils Bohr in Denmark, Hahn and Strassman in Germany and Fermi in Italy.

Most elements are composed of a mixture of normal atoms and isotopes. These are atoms which have a larger or smaller number of neutrons than the norm. Uranium, which was found to be especially suited to fission, has an isotope which is lighter than the norm: it has an atomic weight of 235 as compared with the usual 238. It was discovered that if such a uranium atom were split, several neutrons would be released which then would cause further atomic nuclei to disintegrate and so release more neutrons. This is called the chain reaction.

If such a reaction were permitted to occur very fast the result would be a violent explosion: an atomic bomb. If the rate of fission could be controlled, like banking down a fire and feeding in fuel slowly, this could produce nuclear power to provide electricity for peaceful needs as well as radioactive substances widely used in such fields as medicine, chemistry, archaeology and engineering. As is well known, both means have been perfected. The first atomic bomb was detonated in 1945. The world's first nuclear power station was opened at Calder Hall in Britain in 1956.

The world's first atomic reactor was constructed on the racquets court of the sports stadium of Chicago University and its chain reaction started on 2 December, 1942. Many scientists had taken part in the experiment, but Enrico Fermi (1901–54), the ex-patriate physicist from Rome, was the leader. The experiment was to prove whether a controlled release of energy by nuclear fission was possible. The reactor, called an atomic pile, was built up in layers, the fuel (uranium in the form of metal rods and spheres of uranium oxide) and the moderator (of graphite), whose purpose was to slow down the rate of neutron emission from the uranium. In case the rate of fission became too high and the release of energy too great, cadmium rods, which would absorb the neutrons, could be inserted into the pile to slow the chain reaction or stop it completely.

Fermi's experiment was successful, and a prototype for modern nuclear reactors, which still comprise the basic fuel, moderator and control rods, but it lacked two important factors which were incorporated into later reactors: a biological shield to protect the operators from radiation and a cooling system (42).

In Britain the Calder Hall reactor was followed by two nuclear power stations for the Central Electricity Generating Board: Bradwell (Essex) and Berkeley (Gloucestershire) begun in 1957 (44). These are thermal reactors using natural uranium in sealed rods within a graphite core contained in a steel pressure vessel surrounded by a concrete shield. The coolant is carbon dioxide gas which carries the heat released by the reactor to the heat exchangers, that is boilers which raise steam to turn turbines which then rotate the electric generators (43).

Plutonium is an artificial radioactive element which is produced as a by-product of thermal reactors. For many countries such as Britain uranium has to be imported so the fast reactor was developed which could breed and utilize plutonium so producing greater power and saving new materials by converting more fissile material than it absorbed. A fast reactor can extract from a given quantity of uranium over 50 times as much energy as a thermal reactor. An experimental fast reactor was built at Dounreay in the north of Scotland and began operating in 1959. Today in Britain the advanced gas-cooled reactor has been developed which has higher working temperatures and is more economical and efficient.

There are two methods of releasing nuclear energy. Existing power stations employ the fission means. The alternative is by fusion. This process, of joining together the nuclei of hydrogen atoms, takes place in the centre of the sun at a

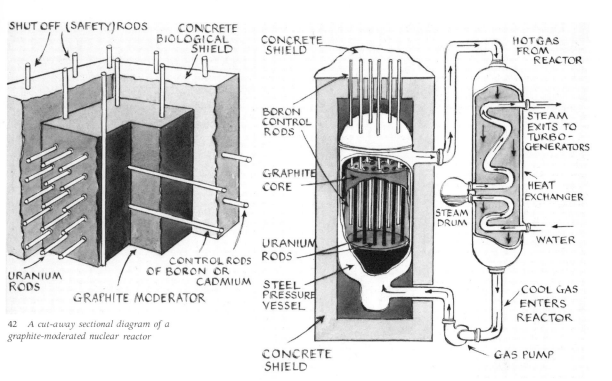

42 *A cut-away sectional diagram of a graphite-moderated nuclear reactor*

43 *A simplified sectioned diagram showing how a gas-cooled nuclear reactor heats carbon dioxide gas to make steam which drives a turbine to generate electricity.*

44 *Berkeley Nuclear Power Station, Gloucestershire, 1959. Note: two reactors each surrounded by eight heat exchangers*

temperature of about 50,000,000°C so transmuting them into helium atoms and releasing great energy. Research has been taking place since the end of the last war to try to simulate this process, which would be an ideal form of energy production as there are ample resources of hydrogen in the world's oceans. An uncontrolled fusion of hydrogen nuclei was achieved in 1952 in the detonation of the hydrogen bomb, but so far attempts to create conditions for a controlled continuous fusion process have been defeated by the problem of creating the high temperature required — about 100,000,000°C. World-wide research programmes continue. Fusion offers the promise of an immense new energy resource by, perhaps, the year 2000.

CHAPTER THREE

Food

In Europe it was about 3000 BC when man emerged from being solely a hunter to begin to till the land to grow crops and to tend herds of domesticated animals. These provided him with meat, milk, wool, fur and leather as well as supplying his ancillary needs in fats, thread, bone and horn. From this time onward food could be produced in adequate quantity for the size of population. In the Middle Ages in Britain this was small; the common land was cultivated in open fields with simple wooden tools and these were sufficient to grow cereals and feed the livestock.

By the time of Elizabeth I the population was steadily increasing and new means were found to supply it. Land enclosure was gradually initiated, making smaller fields which were surrounded by hedges, separating areas for wool production, grazing specially bred cattle or growing particular crops. Farms became larger. Tools and equipment were improved in design and efficiency. The discovery of lands overseas introduced new foods into Britain.

The advances in farming methods were speeded up in the eighteenth century, a time often described as the Agricultural (or Agrarian) Revolution. This was not quite parallel to the industrial one, as it owed more to an empirical approach to agricultural practice than to mechanization, which took place later, in the second half of the nineteenth century. At that time also, scientific research brought a fuller understanding of plant and animal nutritional needs; also the ability to manufacture artificial fertilizers and means of pest control.

More recently, the immense rise of population has led to food technology, a factory production of food based upon a study of chemistry, physics, biology and nutrition, without which processes a country such as Britain would not be able to feed its people.

FARMING METHODS

Under the open-field system practised from before the time of the Norman Conquest the land was divided up for cultivation into narrow strips with grass bands between. It was found by experience that, if a crop was grown in the same soil continuously for a number of years, the results became less fruitful, so the three-field plan evolved. In this wheat and rye were grown in one field, peas, beans, barley and oats in another and the third was left fallow to recover its productive capacity. Such use of the fields was rotated each year and everyone had holdings in all three. While the crops were growing the fields were fenced off. After harvest the fences were removed and the village farmers could graze their animals there as well as on the fallow land. The system was not very efficient but sufficed for the population to be fed.

Each autumn a large part of the herds of livestock was slaughtered because the supply of winter feed was inadequate. The meat was salted and cured in each household for winter consumption. Also, because so many animals were killed, the fields were inadequately manured to produce good crops the following year.

During the sixteenth century the discovery of the New World and the developing trade with the Indies and the Far East led to the introduction into Europe of many hitherto novel foods, spices and drinks, so bringing greater variety to the diet. Crops such as potatoes, maize and rice were grown: sugar and, later, tea, coffee and cocoa were imported. Buckwheat was increasingly cultivated and oil-producing plants such as rapeseed were experimented with.

By the 1650s experience was showing how yields could be increased in order to feed a growing population. Led by the Dutch, then the English, European farming methods were gradually improved. This was achieved partly by enlightened wealthy gentlemen farmers who ploughed back into their estates profits from their commercial undertakings, partly by enclosure and partly by the introduction of new farming practices based on an increased understanding of soil husbandry and plant needs, also the introduction of new crops from abroad and selective breeding of livestock.

During the eighteenth century the process speeded up. It was a time of great expansion with larger yields in crops and animals, in considerable part due to a vastly increased acreage of land under the plough. This increase came from the land previously thought of as waste – forest and moorland areas – as well as that reserved, under medieval custom, as fallow or common land. The medieval pattern of having up to one-third of agricultural land lying fallow to recover from two years' cropping, as well as the open-field system, which meant that crops and livestock of poor or diseased condition were mixed willy-nilly with quality stock, produced a subsistence level of farming which was quite unable to cater for the vastly increased eighteenth-century population.

The answer to the problem was enclosure, which took place, at varying speeds, over hundreds of years from the Middle Ages until the later nineteenth century but which was implemented most quickly between 1760 and 1815. The enclosure of land into separate fields bounded by hedges made it possible to separate farms operated under good husbandry with vigorous

crops and selected beasts from those of poor quality. It enabled larger, more efficient farm units to become established and to permit a rotation of crops which led to higher yields. Many enclosures were carried out by agreement among owners, but a succession of Acts of Parliament were passed to speed up the process. Inevitably, hardship was caused to some smaller farmers and cottagers and there was much bitterness, but many small farmers adapted and prospered and the country benefited extensively from the new system. Without such change starvation would have been widespread, especially during the Napoleonic Wars.

Crop yields were also spectacularly increased with improvements in the soil's texture and fertility by marling, manuring and drainage. Marling was an ancient practice, revived more than once but particularly during the seventeenth and eighteenth centuries. In this the marle (heavy clay and lime in the subsoil) was dug up and mixed with the lighter topsoil, so improving texture and water retention. At the same time a system of summer stall feeding of cattle and the housing of sheep produced accumulations of organic manure, also arrangements were made for town sewage and factory waste to be collected and supplied to farms.

Of vital importance to crop yields were the use of the seed drill to replace broadcast sowing and the horse hoe to keep down weeds (Tull, page 48) and the introduction of the system called the Norfolk Rotation. This was a four-year sequence of crop sowing, usually in the order of wheat, turnips, barley, clover, which had originated in Holland and was then adopted by Townshend in Norfolk. This did away with the old need for one fallow year as the turnips broke up the soil and the clover fed it with nitrogen (though at the time it was not understood how or why it did this).

The growing of root crops such as turnips (introduced from sixteenth-century Holland) and, later, swedes and mangolds, rye grasses, sainfoin and clover-like plants such as trefoil and lucerne as fodder crops made it possible to over-winter larger herds and these in turn provided all-the-year-round organic manure. As the quality and quantity of livestock improved, selective breeding became possible, leading to outstanding breeds of cattle, horses and sheep.

The great eighteenth-century farming pioneers in England did not, in general, originate these new methods but they experimented with and popularized them so that gradually they became common agricultural practice. Most famous among them were Lord Townshend (1674–1738) and Thomas Coke (1752–1842) of Norfolk and Robert Bakewell (1725–95) of Leicestershire.

In 1840 Professor Justus von Liebig (1803–73), German organic chemist, inaugurated the modern approach to plant nutrition in an address delivered to the British Association (45). He pointed out the need of plants for specific inorganic salts and that they required feeding with, among other substances, phosphorus, potassium, lime, magnesia, potash in order to reach maximum growth. Soon Liebig turned his attention to animal nutrition, analysing the chemical composition of the food required for healthy growth. His ideas stimulated scientific research and later in the century results of this work were

being reflected in the careful study of nutritional needs for crops and livestock.

Meanwhile, in 1843, Sir John Bennet Lawes (1814–1900), the English agricultural chemist, established, in conjunction with Sir Henry Gilbert, the Rothamsted Experimental Station in Hertfordshire to make a practical scientific study of crop nutrition. Lawes made a fertilizer by dissolving bones and mineral phosphates in sulphuric acid: a superphosphate. He began to manufacture this on a large scale in 1842 when he founded the world's first chemical fertilizer factory.

By the late nineteenth century, with selective breeding of plants and livestock, scientific nutritional care and soil treatment, improved transport and communication facilities, mechanization powered by steam and the growth of the fertilizer and herbicide industries, it had become possible to provide food for the unprecedented rise in population of Victorian England while, at the same time, a smaller proportion of these people were working on the land. Meanwhile England, which had led the agricultural advance for so long, had given place to the New World. In North America and Australasia the vast areas devoted to the growing of crops and raising of livestock needed new methods and machines and it was from there that many of the innovations in mechanization and scientific breeding stemmed.

During the twentieth century, and especially after 1945, a greater variety of foods became available from the home market and from abroad and a rising standard of living, together with smaller family units, brought a demand for a better diet, notably an increase in consumption of meat, eggs, milk, fish and fresh vegetables. Families wanted smaller, tender joints of meat with less fat than previously. Part of this demand stemmed from researches into nutrition which showed the importance to human and animal diet of protein, fat, carbohydrates and, with the work of Sir Frederick Gowland Hopkins (1861–1947), the English biochemist, and others between 1893 and 1912, the important part played by essential amino-acids which came to be called vitamins.

Rapid advances in animal breeding resulted from intensive studies of genetics, so producing animals with a more desirable proportion of

45　*Justus Liebig, 1843*

tender meat to fat and bone; a young animal richly fed in the early stages of its growth displays such qualities. It was also discovered which breeds of animal were more responsive to such special feeding. Artificial insemination of cattle was a breeding technique successfully developed in the 1920s in the USSR. By the 1930s it was being practised in the West. The discovery in 1949 of the means of satisfactorily freezing semen for extended periods initiated the tremendous expansion in the use of this process after this.

Genetic studies also led to improved quality in other animals (lamb and pork) as well as in poultry and eggs. Similar research developed better crop strains and the astonishing advances in selective pesticides and herbicides gave protection to growing plants. The laboratory synthesis of antiobiotics has proved a vital contribution in combating disease in plants and animals as well as stimulating growth.

MECHANIZATION IN AGRICULTURE

Until the later eighteenth century the tools and equipment used in tillage were simple traditional ones made of wood (with parts of iron by the village blacksmith) by the village carpenter or by the farmer himself. Tillage is the preparing and use of the soil to grow crops. From ancient times this meant ploughing in order to turn over the earth to open it to the air and smother the weeds, then breaking down the clumps by cultivating and harrowing to provide a fine tilth for a seed bed and finally rolling to smooth and pack the soil.

Early ploughs were entirely of wood pushed by one man or pulled and pushed by two. Later, teams of oxen and then horses driven by a ploughman were used (46). In Europe ploughs were generally fitted with a mould-board which turned over the sliced earth, so burying the weeds. Wheeled ploughs were more suitable for heavy soils as they gave support but swing ploughs were in general use at the same time. Between about 1775 and 1850 factory-made ploughs generally replaced local-made ones and cast-iron was used for working parts rather than wood. In the 1780s Robert Ransome, founder of the Ipswich firm of that name, introduced a self-sharpening, cast-iron ploughshare which was hardened by a chilling process. In 1808 the factory was making iron frame ploughs with parts standardized for easy replacement.

Harrows and cultivators were also of ancient origin. These were traditionally of wood, made up into rectangular frames fitted with projecting spikes to score the soil. Iron tines replaced wooden ones in the sixteenth and seventeenth centuries. By 1800 the wooden frames were succeeded by iron ones; a heavier design of cultivator was mounted on wheels. Early rollers were also of wood. Later, the best known innovation which greatly improved crop yields was the effective clod-crusher patented in 1841 by Crosskill. This spiked roller, weighing 26 cwt and needing three horses to draw it, was shown at the Great Exhibition in 1851 (47).

The traditional method of sowing seed was by broadcast, that is, by a sower who carried the seed in a basket supported on a sling over his shoulder and scattered the grain by constant handfuls. Later, the dibble method was used where a man made holes in the ground at measured intervals with a wooden tool and women and children followed him dropping seeds into the holes; this was suitable for large seeds such as those of beans.

Jethro Tull (1674–1741) was the first to produce, in 1701, a satisfactory design of horse-drawn seed drill which cut rows of furrows for the seeds and which sowed them at suitable depths in these (49). He was encouraged to make such a drill by the then high price of seed and his results, in yield, from small seed such as wheat and sainfoin, were startling. Since there was no waste of seed and each was sown in the correct place in the soil, his yield increased eight-fold over the broadcast method. He also designed a horse-hoe to keep the spaces between the rows weed-free: this had been of no use before the seed was sown in rows. Tull published details of his inventions in 1733 in his book *The Horse-Hoing Industry* and before long improvements were being made to his designs; most notable was James Cooke's drill of 1782. Soon drills were being made to sow seeds of all sizes, but it was after 1800 before the use of such drills became general.

46 *Ploughing with horses. W.H. Pyne, 1802*

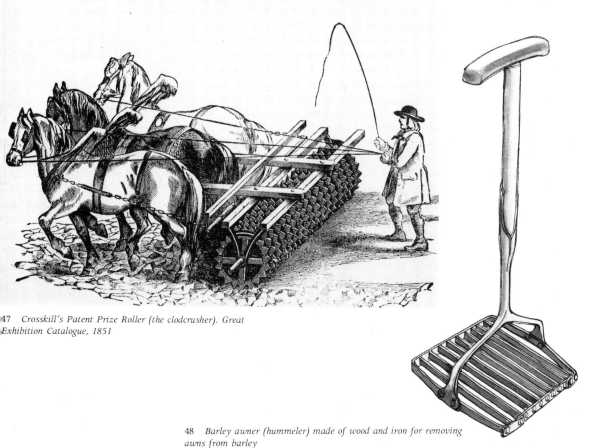

CROSSKILL'S PATENT PRIZE ROLLER.

47 *Crosskill's Patent Prize Roller (the clodcrusher). Great
Exhibition Catalogue, 1851*

48 *Barley awner (hummeler) made of wood and iron for removing
awns from barley*

Food

Before mechanization, corn harvesting processes were more labour-intensive than any other tasks on the farm. The crop was cut by hand with a sickle or scythe. It was then gathered into bundles, bound and stood in stooks to dry in the air. When dry the crop had to be piled into wagons and taken into barns to be threshed. The sheaves were laid out on the threshing floor and the grain was separated from the ears either by the hooves of oxen, mules or horses being driven round and round upon the sheaves (a system still in use in Spain in 1956) or beaten by hand flails. Finally the grain was winnowed, when the chaff was carried off in the wind as the piles of grain were thrown into the air with hand shovels.

Only slowly were these processes mechanized and, though many different machines were devised in the eighteenth century, it was not until well into the nineteenth that successful models were being manufactured. In Scotland, the Reverend Patrick Bell produced the first satisfactory reaper in 1826 (50). This worked on a scissor principle and was pushed by a team of horses. Much more efficient and widely-selling was the American design of Cyrus H. McCormick, invented in 1834 and shown in England at the Great Exhibition, after which it became very popular. This worked on the more modern knife and cutter-bar principle and was pulled not pushed (51). After this Crosskill of Beverley reintroduced a similarly designed revised version of Bell's reaper which was in general use in the 1850s. The combine harvester (the machine which combines reaping with binding and threshing) was patented in America as early as 1828 but was not in successful use until the 1870s when, in California, it was handled by four men driving teams of 24 horses. Its use did not become widespread until animal power could be replaced by that of the steam engine or internal combustion engine (55).

Threshing machines appeared a little earlier, the first being patented as early as 1636. It was followed by many other schemes, most of them based on the flail principle. The first practical example was that designed by the millwright, Andrew Meikle, in Scotland in 1786; this worked in a similar manner to the machine which had recently been developed to scutch flax (page 133), that is, a drum fitted with teeth rotated at speed inside a circular shield called a concave, with only a small clearance between the two surfaces. The grain was fed in between the drum and concave and the husks were rubbed off. This principle was widely adopted and the machine was soon in general use all over Britain. It could be powered by hand, by horse or, later, by steam (11). Winnowing machines were developed by the 1780s. These incorporated a fan and were turned by hand to blow the chaff from the grain. Later models combined the two processes of winnowing and threshing.

Haymaking was not successfully mechanized until after 1856 when an American design of mowing machine became available which was fitted with a flexible cutter-bar to adjust to uneven ground surfaces. In the 1890s a sweep-rake was introduced which gathered up the cut hay, and by the early twentieth century elevators were being designed to raise the hay to the stack by means of a revolving belt.

The process of mechanization was accelerated during the nineteenth century as steam power was applied to farm equipment. Steam engines were adapted to drive barn machinery of all kinds: threshing machines, root cutters, etc. Stationary steam engines were used for ploughing; at first one was usually placed at each side of a field and the plough was pulled backwards and forwards by cable (52). Steam was later used to drive reapers (6).

The great breakthrough in powered farm machinery came with the development of the internal combustion engine. The first tractor was made in America in 1889 and tractors were in use commercially before 1900 though the first successful lightweight farm tractor (run on paraffin) was designed by an Englishman, Dan Albone, in 1902 (53). Few were in use until the First World War, when lack of men and horses on the farm and the blockade by German submarines caused an acute shortage of food. Henry Ford, who by 1917 was mass-producing tractors as he had cars, came to the rescue and supplied Britain with the means to mechanize food production.

Afterwards, the tractor was first used to power existing equipment but, as its potential was

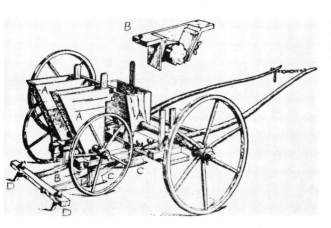

49 Jethro Tull's seed drill, 1701. Seed was fed into hoppers (A). It fell into seed boxes fixed to base of hoppers (B and inset B) and from there into funnels (C). From here seeds were directed into the furrows in the earth made by the machine and the harrows (D) covered them with soil

50 Bell's reaping machine illustrated in Encyclopaedia of Agriculture J.C. Loudon, 1831

51 (Below) McCormick's patent reaping machine from Ransome's catalogue, 1859

52   *Cable ploughing by steam engine*

53   *Dan Albone's Ivel Agricultural Motor (tractor),*
1902. From Implement Manufacturers' Review

54   *Agricultural tractor, 1981*

55   *Combine harvester, 1981*

realized, new machines were designed specifically for work with tractors and to combine various operations as in the case of the combine harvester. By 1950 the tractor had almost replaced the horse; its increased haulage power and manoeuvrability brought an immense increase in food production (54).

Although efforts in America to find a way of milking cows mechanically began as early as 1819, these were mainly on the siphon principle and were both unhygienic and harmful to the cow. Experiments were then carried out to try to simulate the sucking action of a calf, and the British patent of 1851 was an attempt to use vacuum for this. However, it was after 1900 before successful machines were operating. Internal combustion engines were used at first to power the machines, then electricity as the supply was laid on at farms (56). In Britain progress was slow and machine milking only became general after the Second World War. Another important advance in dairying was the invention, in 1877, by the Swedish engineer Gustav de Laval of the centrifugal cream separator. This made immense economies possible in labour and space because the dairy no longer had to be filled with the large shallow pans in which the milk was set waiting for the cream to rise (58).

PRESERVATION OF FOOD

Of the six principal ways to preserve fresh food — addition of chemical preservatives, drying, salting, smoking, chilling and canning — only the last does not trace its origins back to classical times. The traditional use of chemical preservatives was generally in the form of vinegar, salt and vinegar or sugar. With sugar, in the case of preserving wines and ales, a process of alcoholic fermentation took place. In the nineteenth century substances such as borax, formalin and sulphuric acid were found to inhibit the acitivity of micro-organisms, but their use was later forbidden when it was shown to be harmful to humans.

Salting was used by the ancients as a means of preservation of both fish and meat. By experience the Romans developed scientific methods of salting where they quickly salted fish caught in the Mediterranean Sea and transported it to all parts of the Empire. In the Middle Ages merchants of the Hanseatic League organised large-scale operations in the North and Baltic Seas, speedily gutting the fish after landing, then, having washed it in sea water, packed it in wooden barrels with salt and brine. Carefully treated, the fish would keep for at least a year. At home families cured meat by salting methods to keep it all winter.

Smoking fish was an alternative, and ancient, method. There were several ways of doing this, some by cooking and smoking, others by pickling and smoking. All kinds of substances such as oak chips, peat or sawdust were burnt to give colour and flavour to the food.

All three means of preservation — chemical, salting and smoking — are still in use. Modern technology has brought new ways of treating the food, as with the curing of bacon and ham, but the advent of the twentieth-century methods of dehydration, canning and freezing has greatly reduced the use of these earlier methods. Drying the meat and fish is probably the oldest method of preserving food; sun-drying where the climate was suitable, air-drying elsewhere. From the 1870s experiments were made with dehydration, drying the food artificially. Several ways of doing this have been developed in the twentieth century, especially during the Second World War when spray-drying was widely used. Liquid eggs and milk were sprayed into a vessel through which hot air was blown and this caused the droplets to dry and become powder. Since then other spray-dried foods such as soups, coffee and mashed potato have become available. Freeze-drying is a process in which the food is first frozen, then dried under vacuum (that is, in a chamber from which most of the air has been removed) until the ice has been evaporated. The process was known in the 1930s, but until the 1950s when a large-scale plant for freeze-drying food was built in the USSR, it was chiefly used in the field of medicine. The great advantage of freeze-dried food is that its flavour and aroma are well retained and it reconstitutes excellently. The disadvantage is its cost; it requires elaborate equipment with skilled operatives and takes a long time. In Britain its chief use is in coffee granules. Other modern forms of dehydration include puff-drying, used especially for

56 *The Thistle milking machine in use.* Royal Agricultural Society Journal, *1895. This mechanism, patented by Alexander Shields in 1895, introduced the pulsating action into milking machines. This was a very important innovation and, though this particular machine was not successful, improved designs became available within a few years*

57 *Ice (refrigerating) cabinet in 'Harrods' Catalogue, 1929. Wooden cabinet, insulated and lined with steel*

58 *Mid-eighteenth century dairy. Setting milk and making butter*

vegetables and foam-mat drying which has been found to be particularly suited to baby foods, eggs, milk and concentrated fruit juices.

Milk has always presented special problems since, untreated, it goes bad so quickly. When Britain was still an agricultural country, getting milk from the cow to the customer was not difficult as towns and villages were small and a cart, or even the cow, could be walked on the milk round. With industrialization towns became large and fast distribution a problem. The railways helped, as did the metal containers which delivered the milk to dairymen but it was pasteurization which made the milk safer. Louis Pasteur (1822–95), after whom the process was named, published his paper on bacteria and fermentation in 1857, but it was another 50 years before commercial dairies were pasteurizing milk and in Britain the process was not readily accepted until much later. Pasteur had shown that heating liquid food would kill most of the bacteria which would turn it sour. The milk was heated to a temperature of between 62°C and 65°C for 30 minutes. Later, after 1940, the higher temperature system has been more generally adopted, to heat milk to 72°C for 15 seconds then cool it quickly.

In the mid-nineteenth century other ways were found of selling milk which would keep. Evaporated milk became available; in this process part of the water had been removed by evaporation and the product of reduced quantity could be sold in cans. Condensed milk, which was partially evaporated, and with sugar added as a preservative, was also sold by 1856. Nowadays Long-life milk can be purchased which will keep at home for six months. This is fresh milk heated to 130°C for one second.

It was Nicholas Appert (1749–1841), the Parisian confectioner, who first successfully experimented with heating food in sealed containers in order to preserve and make it transportable. In 1809 he received a grant of 12,000 francs from Napoleon to publish his invention, which he did the following year. Appert used glass jars, but in 1810 Peter Durand in England had the idea of using iron canisters. By the 1830s canned foods were being supplied in Britain to the armed services, but over the next 30 years there

were a number of containers where the food went bad or they 'blew'. Then in the 1860s, following upon researches by several scientists, Pasteur among them, a better understanding was gained of the scientific principles involved in this method of food preservation. More care was taken in filling the cans and heating them to sufficiently high temperatures. By the 1870s reliable autoclaves (pressure vessels for heating) were in use for this purpose.

Soon canning became a very important industry in North and South America and Australasia in order to export quantities of food to Europe. In more recent years great improvements have taken place in the industry in methods of heating the cans, in metals used and their coatings, in sterilization, sealing and automatic means of production.

Since very early times it has been known that fresh food will keep longer if the temperature is lowered, and cold water, ice and snow have been used for this when available. In Britain from the Middle Ages onwards at large houses, castles and monasteries storage places have been built (usually underground) for ice which was collected from lakes and rivers in winter-time.

With the urbanization of the nineteenth century the demand for ice grew quickly. Farmers flooded their fields in winter and sold the ice to great estates, trawler fleets and the insulated port warehouses being built for storage purposes. From the 1830s ice was imported in quantity from the USA and Norway and soon ice boxes were being sold for the home. These were insulated wooden cupboards lined with zinc or slate in which the ice, delivered daily, would keep the food fresh (57).

Many attempts were made to make ice artificially. Modern commercial refrigeration is largely based upon the work of the German professor of thermodynamics, Karl von Linde (1842–1934), who introduced an ammonia-vapour compressor system in 1876 (ammonia has a critical temperature of 132.4°C, well above that of the room). This led to a wide-scale development of plant which came to be used by the 1890s for carrying food cargoes on long voyages by sea.

The household mechanical refrigerator was marketed in the early twentieth century. The compressor type was driven by a motor which was at first powered by steam, then gas, but with the development of the small electric motor (page 39), more commonly by electricity. Most of the research and development of the domestic refrigerator was carried out in America. Britain lagged behind in acceptance, partly because of the cooler climate, so that in the inter-war years it remained a luxury article (59). After the Second World War, sales increased and by 1960 half the population had access to one. Since then the home freezer has also become popular to store frozen foods either purchased at the supermarket or prepared at home.

The chilled and frozen food industry which has revolutionized shopping and the preparation of food for the British housewife since 1950 also originated in America. The chief pioneer was Clarence Birdseye (his famous firm still bears his name), who set up his first fish freezing plant in New York in 1923. It was Birdseye who, in 1929, realized the importance of the fact that rapid freezing of many foods to a temperature of −23°C or lower retained the flavour and texture much better. After the war, plants were set up in Britain which for some time were devoted to freezing fish and peas as these two foods gave such satisfactory results: soon fish fingers were born. Before long experience gained led to successful freezing of a wide range of meat, fish, vegetables and fruit.

FOOD PROCESSING, MARKETING AND PACKAGING

During the twentieth century, and particularly since 1945, the production and distribution of food has completely changed. The industry has expanded beyond recognition to provide a product which is hygienically safe, with its nutritive value scientifically balanced; all harmful attributes have been removed and, many would say, much of the flavour also. The industry is no longer merely supplying an agricultural product but one which emanates from a complex system of food technology. There are many advantages to this type of food. Apart from health and safety, it is conveniently packaged and attractively presented, it is easy to store and to purchase, it is fresh and, in the case of convenience foods, saves

the housewife time and trouble. Farming and horticultural products are mostly seasonal, but freezing processes and aircraft transport mean that many products are available all the year round. There are also disadvantages. Flavour and character seem to have been diminished and fewer varieties of a given food are available; this is particularly apparent in strains of fruit and vegetables. It is not easy to purchase a food which is not generally popular. But, whatever the balance sheet on food technology, it is certain that, without scientific, sophisticated farming and food production methods, the present size of population could not be fed.

Chemical additives to food have been in use since early times, for example, salt as a preservative, but modern food contains additives of a much more varied kind and for a multitude of purposes. These include agents for colouring, flavouring, sweetening and bleaching; nutritional supplements such as vitamins, preservatives and, for health safety, antibiotics and insecticides. Much controversy attaches to some of these additives as, for instance, colouring and flavouring agents, but human beings are conservative about their food: if butter is yellow margarine should be also.

In Britain the booming market in convenience foods has grown up as a by-product of the frozen-food industry as these dishes are chilled or frozen. Although most housewives would not plan their family's meals entirely on such foods (they would be too expensive anyway), few have not turned to them in relief in emergency or in part answer to the daily problem of 'what shall I give them for dinner tonight?' Such foods range from the ready-prepared steak pie, quiche or pizza which only need heating, to a kebab of meat and vegetables on a skewer requiring cooking, and a complete television meal which can be presented almost without taking one's eyes from the screen.

The production of structured foods from sources not fully tapped previously is a more recent development and provides a cheaper food than that upon which it is based. Methods have been found to use cheap, tough, fat and gristly meat to produce acceptable products by heating or compression. Imitation meat fibre has been produced from soya beans (so-called 'knitted steaks') and is

59 *Electrolux water-cooled, absorption refrigerator, 1927. Wooden cabinet insulated with cork and lined with painted sheet metal. This model originally supplied to King George V and used at Sandringham until the early 1950s*

60 *Convenience foods. Findus beefburger production line – entirely automatic production and one of the fastest in Europe. The ingredients are fed into a blender and burgers are stamped out from the mixture. They are then passed through a freezer and, after this, a packing machine. They are kept in cold storage until distribution to shops*

widely used in meat products such as sausages or meat loaf. Research also continues into producing a greater quantity of food by synthesis from sources hitherto not thought of as food, for example, algae or petroleum.

Margarine was an early product of food technology. It was invented in response to a prize offered by the French government for a cheaper substance in imitation of butter for use by the army and the poor. Production began near Paris in 1870 of a method produced by Hippolyte Mège-Mouriés in which animal fats were rendered down and emulsified with milk and colouring matter. In England the product was called butterine, but from 1 January 1888 the term was outlawed when it was decreed that it should be sold as margarine.

Soon after 1900 it became possible to use vegetable oils to make margarine instead of the more strongly flavoured animal fats. This resulted from the discovery of the catalytic hydrogenation process in which hydrogen could be introduced into the molecules of the vegetable oils thus altering their chemical composition to give them a higher melting point and so hardened them to a suitable consistency for margarine. Today a wide variety of vegetable oils is used in production with colouring and flavouring agents added for palatability.

Among the many fast-growing industries of the twentieth century which are dependent on food technology are all kinds of breakfast cereals and a variety of potato chips and crisps. Breakfast cereals originated at Battle Creek, Michigan in America, a result of the nineteenth-century religious interest in health foods. Dr John Harvey Kellogg invented several cereal and nut foods, including peanut butter and flaked breakfast foods for his patients in the local sanatorium. It was his younger brother, W.K. Kellogg, who developed the cereals for world-wide sales. Corn Flakes and All-Bran came to England in 1924, soon to be followed by Rice Crispies.

Potato crisps were first made by a Frenchman called Cartier. He came to England in 1914 and five years later sold his business to Frank Smith who set up a tiny factory in Cricklewood to fry sliced potatoes. The crisps were packed in waxed paper bags and these, not being airtight became soggy after 24 hours. It was after 1946 that the crips industry 'took off', with several producers and a complex distribution system. Packaging is now in plastic film; sliced potatoes have been replaced by potato flour.

Factory farming is a controversial subject. It embraces eggs and poultry, also calves and pigs. Some critics of the system abhor this means of high productivity while others complain that the eggs and meat are tasteless; so there is a market for the limited free range system but sufficient quantity could not be produced by using only the older methods.

In Britain the poultry industry is big business and methods are intensive and comprehensive. Farms comprise units for laying, breeding, incubating, hatching, rearing, processing and marketing. Laying birds spend their 52-week lives in individual cages in a completely controlled environment of heating, lighting and graded feeding to give optimum egg production. Table birds are similarly reared though running free in large broiler houses and at ten weeks old are passed through an automatic conveyor system to be successively electrically stunned, bled, scalded, plucked, eviscerated and decapitated, trussed, vacuum packed in plastic film and finally frozen to be 'oven ready'.

During the last 50 years the pattern of marketing and packaging of food has been transformed. The two systems are interdependent. With the advent of self-service shopping in the supermarket it became essential, both for hygiene and convenience, for the food to be weighed, priced and packaged and the development of the plastics industry has made this much easier. Equally, the growth of the large retailing groups which opened the supermarkets needed to be able to make heavy investment and this was only possible with changed methods of selling.

The chain stores which were built up to serve the mass consumer food market were being established by 1870: Sainsbury and Lipton, then Home and Colonial 1885 and Tesco 1930. Successful chains bought wisely, kept costs and prices low, gave good value, had a high standard of hygiene and cleanliness and offered good service. Self-service stores began in America in 1912; in Britain the first experiment was in 1942, but food rationing restricted this type of selling until the 1950s.

Before 1930 packaging was provided by paper, card, tinfoil, wood and tinplate. Cellophane, a moisture-proof cellulose film, was widely used from 1937 when production began in the Bridgewater factory. Since the 1960s, as more and varied plastics became available (page 83), these supply packaging materials in use for wrapping, padding and food containers of all kinds.

COOKING METHODS AND FUELS

The prime factor which has dictated the design and means of domestic cooking over the centuries has been the fuel available at the time. During many hundreds of years, from primitive times until the seventeenth century, this fuel was chiefly wood which was burnt in an open fire upon a brick or stone flat hearth. At first in the centre of the room, soon it became a wall fireplace set in a deep arched recess with a wide flue above. Logs were supported on andirons and the wall was protected by a cast-iron fireback.

Food was cooked over and in front of this fire. For roasting, meat, poultry and fish were turned on a spit in front of the hearth and the fat dripped into a tin beneath. Various means were developed to turn this metal spit, from the early boy turnspits, then dog turnspits in wall cages, to gravity and mechanical clockwork designs (61), and, finally, the sophisticated smoke jacks in which the spit was turned by means of a chimney vane mechanism powered by the draught. Ancillary aids included bottle jacks for turning small joints (62) and metal hasteners for speeding up the roasting process (67). Various types of grillers and toasters stood in front of the fire (61).

Food was cooked over the fire in metal cauldrons and pans and, over the centuries, means were developed by the use of the chimney crane to support these in the correct position to be maintained at the desired temperature. This comprised an iron post fixed into the floor of the hearth at the side of the fire. A horizontal bar extended from it and from this could be suspended the vessels (61, 62). Ovens for baking were built into the wall or floor by the hearth (66).

Due to the shortage of timber, coal gradually replaced wood as the domestic cooking fuel during the seventeenth century. Coal needs a good draught to burn well and it was difficult to kindle on a flat hearth so the dog grate evolved. This was an iron basket standing on legs on the hearth with the andirons attached at the sides. Soon followed the basket grate, complete with fireback (61). The final stage in the later eighteenth century was the hob grate in which the fireplace aperture was decreased and the firebox enclosed by iron plates at the sides and top, so making the fire burn hotter and providing horizontal surfaces on which to heat kettles and pans (62).

By the late eighteenth century, research and experiment led to a greater understanding of heat conservation and how to design more efficient means of cooking with coal (63, 64). Slowly, the idea that further enclosure of the fire and better circulation of air would lead to economy of fuel and give better heat control was put into practice. The nineteenth-century kitchen range was the logical result. Thomas Robinson designed the first such range in 1780 and this was improved upon by George Bodley, an Exeter iron-founder, who patented his enclosed kitchen range in 1802.

By mid-century there were two main designs of range, one more enclosed than the other. The open one, of which the cottage range and the Yorkshire range were typical, was more popular in the north of Britain because it cooked the food and warmed the kitchen at the same time, but it was extravagant in its consumption of fuel. In use from about 1800 until the 1920s, it had an opened, barred firebox, some bars being hinged to be able to support pans or kettles. The oven flanked the fire on one side, a hot water boiler or warm closet on the other (65). The closed range or kitchener, which became available about 1840, was much more economical of fuel. It was designed to cook or heat, but not both simultaneously. The firebox was enclosed by a metal hot-plate on top and a door in front and when cooking both were closed. To warm the room they could be opened (68).

By 1900 ranges were more efficient and complex, some larger examples comprising two ovens, several hobs, water boilers and warming compartments for dishes, but they were still

61 Brick fireplace with wood surround c.1680–1720. Based upon that displayed in the Hearth Gallery of Castle Museum, York. Note: basket spit rack on chimney breast and, left to right, frying pan, salamander, brass cauldron, iron and steel basket grate, cobirons, basket spit, dripping tray and down-hearth toaster. Spit driven by mechanical weight jack. Chimney crane fixed to back of hearth carries copper kettle with tilter. Wafering iron, ladles, iron cauldron. Note especially salt container built into wall of chimney (left) to keep salt dry, with hole for hand to extract salt

62 Cast iron hob grate, late eighteenth century. Drawn from a combination of museum sources. Brick hearth and chimney. Iron chimney crane and griddle. Steel and brass fender with fire irons. Brass bottle jack. Copper kettle. Note also: salamander, rushlight container (attached to brick walling), marram grass besom, teapot and trivet

64 *Kitchen and cooking range in the house of Baron de Lerchenfeld in Munich designed by Count Rumford in the late eighteenth century. From Rumford's original designs*

63 *Count Rumford (Sir Benjamin Thompson, 1753–1814) whose studies of the motion of heated gases led him to consider the problems of efficient use and conservation of heat in domestic cooking stoves*

made of iron, brass and steel, were difficult and hard work to clean and needed constant black-leading and polishing (68).

As the supply of coal gas was extended in the 1820s (pages 117 and 120), experiments were made to use it for cooking, but the problems were formidable. Gas was much dearer than coal, there were no meters, it was difficult to measure consumption. Hardest to counter was the public prejudice against gas; it was generally believed to be dangerously explosive and the fumes would affect the food if used for cooking. Still, inventors kept on trying. The first workable appliance dates from 1824 when a griller was made at the Aetna Ironworks. In the 1830s James Sharp (1790–1870), assistant manager of the Northampton Gas Company, demonstrated gas cooking in his home and designed cookers which were sold commercially.

Domestic cooking by gas finally began to come of age in the 1850s when several designs of cooker became available, all black cast-iron boxes standing on four legs and containing an oven, grill and hot plate with burners, but it was another 20 years before such cookers became competitive with coal ranges. Prejudice and fear

remained and cost was still high. Improvements followed in the 1860s and 1870s: the introduction of the new bunsen-type burners, the independent gas ring, and multi-burner hot-plate and, most important, availability of cookers to rent and the one penny pre-payment slot machine. These two factors brought gas cooking within range of most people (72).

Not until after the First World War was the cast-iron cooker gradually replaced by the easy-care coloured enamel panelled one, and, in 1923 came that major innovation, the Regulo thermo-static control (70). Since 1950 improvements have quickly followed one another: the high-level, then the Sola grill, automatic ignition, stream-lined shaping, rotisseries, automatic oven-timing, graduated simmer control and the oven flame-failure device.

Electricity, like gas, was utilized for lighting several decades before appliances were designed and manufactured for cooking (pages 117 and 122). Between 1900 and 1914 a number of models of electric cooker appeared on the market, but few people bought them (71). There were a number of circumstances mitigating against choosing elect-ricity for home cooking: only a proportion of

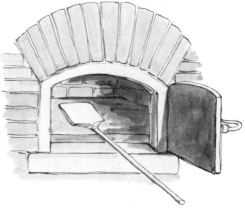

65 *Working family's cottage hearth. Painted wood chimneypiece with brass rail under mantelpiece. Decorative cast iron range with brass fittings. Open firebox and brick backing. Open range comprises one oven, hot water boiler with tap and warming cupboard. Steel fender and fireirons, 1880s. Derived from several similar cottage hearths*

66 *Brick oven with iron door built as part of a hearth at the side of a hob grate. Early nineteenth century*

67 *(Below) Metal hastener (roasting screen), with brass bottle jack*

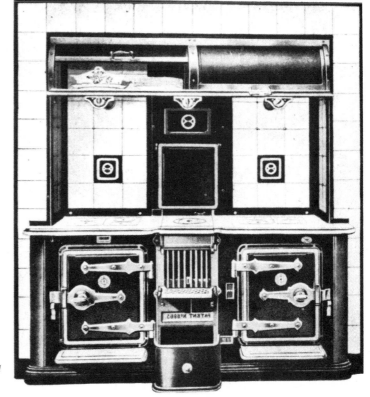

68 *(Right) Patent Quadrant Kitchener c. 1905–10 (catalogue). Cast iron with bright steel fittings, tiled back. Double oven. Hot closets with roll doors*

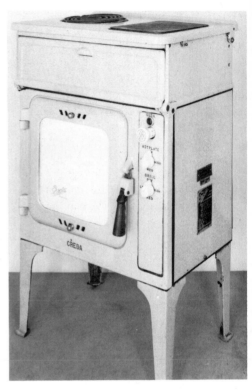

69 (Right) Creda electric cooker, 1933, the first made to have thermostatic control of oven. Hot plate has two elements, one of the new spiral tube type, the other a metal plate with coiled wire element beneath to heat the grill

70 Radiation 'New World' gas cooker, 1923, the first cooker to incorporate 'Regulo'

72 Parkinson cooker c.1890. Made of cast iron with silicate packing for insulation. Water heater at left side. Plate-warming hood over hot plate

71 Tricity electric cooking, 1917

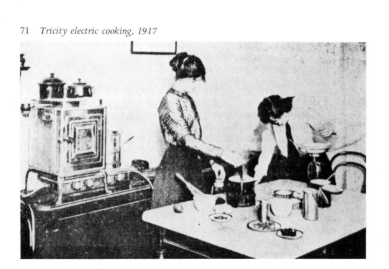

homes were yet wired for electricity; the cookers (of cast iron) were no more attractive or easy to clean than gas or solid fuel appliances; people distrusted a heat for cooking which they could not see (it was only too easy to be burnt accidentally).

Between 1918 and 1939 there were a number of advances which (by 1939) induced one million households to change to electric cooking. Smaller, more convenient cookers were designed by firms such as Tricity and Belling. The hob burners were greatly improved. In early cookers the heating elements were made of wire coiled round cylindrical ceramic formers. As these were open and unprotected, a saucepan which boiled over would often cause the element to short circuit. This type was followed by the solid cast-iron plate burner which was reliable and easy to clean but slow to heat up and cool down, so wasted time and electricity. The fast-heating tubular-sheathed radiant rings introduced in the 1930s were a great advance. In 1933, in belated response to the gas 'Regulo', Creda introduced automatic oven temperature control (69). Since 1950 the design and facilities of electric cookers have marched in step with gas models and now 41 per cent of the British nation cooks by electricity.

Microwave cooking was introduced into Britain in 1959. This is a completely different method from previous ones and makes use of very high frequency radio waves (micro-waves), which penetrate to the centre of the food and, by agitating its molecules, raise its temperature (Chapter Nine).

CHAPTER FOUR

The Home 1
The Materials in Use

For centuries, traditional, natural materials such as wood, earthenware, glass, linen, wool and fur provided the basis for all the requirements of living at home from room coverings and decoration to utensils, means of cooking, lighting and heating as well as home-spun, -woven and -dyed garments worn by all the family. The development in Britain of the major industries of the Industrial Revolution set in train far-searching changes in the materials available to be manufactured into a wide range of goods not dreamt of before 1750. First the iron and steel industry, then engineering and, during the nineteenth and twentieth centuries, most influential of all, the chemical industry, brought advances which affected everything in use in the average home of such widely differing substances as detergents, textiles, matches, paint, plastics and soft drinks.

IRON AND STEEL

Pure iron is one of the elements. It is a metal, silver-white in colour, rarely existing free in nature and of little practical utility, but alloyed, particularly with carbon, it has for centuries been one of the most useful materials. It is found in ores combined with other elements, chiefly oxygen. Such oxides occur in abundance in many parts of the world including Britain.

There are three chief forms in which iron has been used. Wrought iron, which is fairly pure, has, since early times, been used structurally and decoratively. It was hammered while red hot into the desired shape but it could not be sharpened to give a good cutting edge. Cast iron, which

contains a higher proportion of carbon, is hard and brittle so cannot be shaped by hammering but it may be melted and poured into shaped moulds. The most useful and versatile form of iron is steel which is an alloy of carbon and other elements. Many different kinds of steel are made today, the properties required being produced by diverse heat treatments and alloying it with a range of elements to make it harder, tougher, sharper or stainless.

The smelting process had been developed in Asia and came to Europe by 1200 BC. Smelting is extracting iron by fusing or melting the ore. At first, furnaces were simply holes in the ground in which charcoal was burned with the iron ore (its combustion aided by air from bellows), producing a semi-melted metallic lump called a 'bloom'. This bloom was then hammered to beat out the slag or impurities, so making wrought iron. In the late fourteenth century, water power was utilized (page 24) to drive the furnace bellows and power the hammers. Also, the tall blast furnace was developed. In this the iron took longer to penetrate to the bottom of the furnace and, in so doing, absorbed more carbon from the charcoal. The higher carbon content rendered its melting point some 350°C lower than that of pure iron (1535°C) so, for the first time, it became possible to liquidize the metal so that it could be poured into moulds. This cast iron lacked the tensile strength of wrought iron but was strong in compression.

A small quantity of steel was being produced in Europe by the cementation process. This involved sealing iron bars with charcoal in containers and subjecting them to prolonged, intense heat. The iron absorbed carbon from the charcoal and became 'blister' steel. A better quality carbon steel was imported from southern India during and after Roman times. Called 'wootz'* steel it was made by heating iron granules and wood in clay crucibles.

Iron-making in the eighteenth and nineteenth centuries

By 1600, in Britain, demand for iron for firebacks, cooking utensils, farm equipment, etc. was increasing, whereas timber supplies were diminishing, so coal was used more often to smelt iron, but this caused severe problems. The iron was brittle and lacked strength due to the sulphur content of the coal. In 1709 the solution, which was to use coked coal, was produced by Abraham Darby (1677–1717) in Coalbrookdale. Darby's experimentation led to a country-wide movement in the areas in which iron was smelted. Previously, iron-producing districts had been where wood, iron ore and water for power were readily available – notably Sussex and the Forest of Dean. Eighteenth-century development needed access to coal and iron so the new industrial areas centred on Shropshire, South Wales, Scotland, Teeside and Tyneside. Two further innovations made the production of iron cheaper and easier. John Wilkinson, the great ironmaster (1728–1808), installed a Boulton and Watt steam engine (page 30) in his Shropshire works in 1776, so improving the power potential, and Henry Cort (1740–1800) in 1784 introduced the action of puddling. This enabled the cast-iron 'pigs' produced in coking furnaces to be rolled into bars suited for making wrought iron. Puddling was a process of stirring the molten iron to free it of impurities; this was necessary as iron made with coke contained such a large number of impurities that it became brittle and crumbled when hammered.

The nineteenth century was the age of iron. This strong, durable material was employed to manufacture almost anything (73) from its heavy

duty use in building structure (pages 98, 99) and railways to functional kitchen ranges (page 60) and decorative lamp standards and furniture. Ways were found to make a greater quantity of cast and wrought iron more cheaply. Hammering and rolling became powered processes and the introduction of the hot-air blast furnace from 1830 economized on fuel. From the 1860s onwards it became possible to make steel on a large scale so that it gradually replaced wrought iron as the principal engineering material.

Steel-making in the nineteenth and twentieth centuries

Steel had been made in small quantities until the early eighteenth century in Europe by the cementation process (page 66) and from 1740 by the wootz method, introduced commercially into England by Benjamin Huntsman (1704–76) who re-melted the blister steel in closed crucibles to produce a steel with a better, more uniform carbon content. Steel contains more carbon than wrought iron but less than cast iron. In order to make it more cheaply a way had to be found to remove the excess carbon and impurities from the molten pig iron, with a smaller quantity of fuel. It was Henry Bessemer (1813–98) in England (also William Kelly in America) who revolutionized steel-making with the discovery that this could be done by blowing a stream of air through it to create a higher temperature. Bessemer designed his converter (1856), a vessel in which this process was carried out and which could then be tilted in order to pour out the molten metal into ingots (75). Meanwhile Frederick Siemens (1826–1904) was developing, with his brother William, his idea of heat regeneration in which hot waste gases could be utilized to pre-heat the fuel and air entering the furnace. At first solid fuel was used, but William Siemens invented the gas-producer in 1861; this converted the solid fuel into gas which made it possible to obtain a high working temperature of 1,650°C from the use of low-grade coal as well as utilizing scrap iron. These advantages led by 1870 to the growth of this open-hearth method of steel-making as an alternative to the Bessemer process and, by 1900, it had become the principal way of making steel (76).

---

* Though to be a corruption of the Canarese (West Indian) word for steel.

73   Omega decorative cast-iron gas heater, 1900. A small,
portable convector heater containing a bunsen burner

74   English iron domestic clock.
John Holloway, 1611

75   Bessemer converter steel plant, Ebbw Vale, 1860

Charging the liquid iron into the steel furnace

76   *Open-hearth furnace. Pouring molten iron from ladle*

Since the 1920s steel has been used ever more widely in all walks of life. In particular, sheet steel produced in rolling mills fulfils all kinds of needs from trains and cars to office equipment and canning of food (page 56). In the home, it is essential for such mass-produced articles as cookers, refrigerators and washing machines. Most of the steel continued to be made by the open-hearth method until after 1945, but this has now been superseded by the oxygen lance process where pure oxygen is injected into molten steel through a water-cooled lance inserted into the mouth of the converter vessel. Another method (first developed commercially in 1900 in France) using an arc furnace, where an electric arc is struck between electrodes to melt the metal, produces about 20 per cent of British steel.

Today a wide variety of alloy steels are available, some being especially hard for use as cutting tools, some having great tensile strength for building and engineering structures, some needing to be heat-resistant or magnetic or stainless. Such alloy steels are made by adding specific amounts of such elements as chromium, tungsten, vanadium, manganese, nickel, silicon or molybdenum, during manufacture. Stainless steel, for example, contains chromium and, sometimes, nickel. At first, such alloys were produced accidentally, though Faraday (page 35) prepared alloy steels in laboratory experiments as early as 1819, but commercial development only came with the burgeoning production of steel in the later nineteenth century.

NON-FERROUS METALS
Knowledge of metalworking was gained long before the mining of ores began. Several metals – gold and copper, for example, as well as iron in meteorites – were found occurring on the surface of the planet, but soon these supplies became exhausted and it was necessary to mine the ores in which the metals were contained, then, as in the case of iron (page 66), to smelt them. The problems encountered were similar to those in coal-mining (page 26) and iron smelting. Several ores were successfully smelted earlier than those of iron probably because the melting points were lower. For instance, that of copper is 1,083°C,

452°C below that of iron.

Copper was first mined in Asia Minor nearly 6,000 years ago. It is believed that the discovery of bronze (an alloy of tin and copper), the first metal alloy made by man, was accidental, the result of smelting copper ore which contained tin as an impurity. This was about 3500 BC. A millennium was to pass before bronze came into everyday use but by that time it was being cast into all kinds of articles from weapons to cooking pots (77, 81). Brass, also a copper-based alloy, was made by a cementation process (using a zinc ore and charcoal) known in ancient Greece (80). Tin and tin alloys were also employed from classical times for plating copper and iron (79, 82). Probably that most widely used for a variety of domestic utensils was pewter, a generic term for a range of alloys in which tin is the main component. Ordinary pewter, made into larger common vessels, contained four parts of tin to one of lead; best quality pewter had a higher tin content hardened with small quantities of copper, brass and, from c.1600, antimony or bismuth. Lead was an important metal in the building industry, particularly for roofing and piping; it was poured into moulds or made into sheets on boards.

From the 1780s onwards, scientific research into the properties and behaviour of metals and the development of engineering techniques led to an immense expansion of the mining industry. This combination resulted in vastly increased production of metallic ores as well as the discovery of new metals. Among these, most of which were only utilized much later, were platinum, cadmium, nickel, cobalt, manganese, tungsten, titanium, uranium, osmium and molybdenum. In the twentieth century several of these became important in the making of alloy steels, others, together with previously known metals, became necessary in the radio (page 162) and electrical industries, for drilling in the oil industry and in mining, in the explosives field and a myriad other requirements. One of the most important developments, not least in the home, has been the widespread production of aluminium which, although the third most common element in the world, does not occur in nature as a metal but in various compounds, so could only be

77   Bronze ewer, c.1400

78   Silver cap and cover, 1658–9

79   Tin casting (from De re metallica, Agricola, 1556)

80   Brass trivet with wooden handle

81   *Copper smelting by the wash process at St Helens, Newton*
Keates Works. Nineteenth-century engraving

82   *Kitchen utensils being made by the tinsmith and displayed in*
*his shop in France, 1763 (*Diderot's Encyclopaedia*)*

commercially manufactured after the development in the 1880s of the successful electrolytic process of reduction.

Gold- and silver-working were ancient crafts and artefacts of beauty had been created over the centuries to adorn the person and the home (78). Since ancient times both gold and silver had been the chief metals used (with the later incorporation of copper then nickel), in the national coinage. After Von Liebig's (page 47) discovery in 1835 of a chemical method of depositing metallic silver on glass, this silvering process for mirrors replaced the older and more lengthy way of backing the glass to give reflective properties by tin foil amalgamated with mercury. The silver nitrate used for this which was prepared from silver dissolved in nitric acid, was later to become of great importance in the photographic industry in producing photosensitive emulsions for films and papers. As time passed these materials became even more costly and smiths endeavoured to find new methods of plating which would economize in the precious metals. In 1742, Thomas Bolsover, a Sheffield cutler, observed, by accident, that with the application of heat a silver article had become fused with copper. He developed the process which became known as Old Sheffield Plate, to coat buttons with silver. In the 1750s, Joseph Hancock perfected a method of rolling together the heated copper and silver to plate household articles – candlesticks, coffee and tea pots. A decade later, the process was taken up by Matthew Boulton (page 30) and others.

It was the Elkington brothers who patented a system of electro-plating in 1840 whereby gold and silver were used to plate base metals such as copper and brass. In this method, which was based upon Michael Faraday's (1791–1867) laws of electrolysis enunciated by him in 1832 and 1833 (page 35), an electric current was passed from a battery through an electrolyte (which is a solution of a silver salt). The copper article is made the cathode in this solution and silver is deposited from the solution on to the cathode. The electrolytic process is still used, not only for gold and silver but also to plate steel cans with tin for food canning (page 56) and to coat steel with chromium and other metals such as rhodium, for protection, decoration and reflecting characteristics.

THE ENGINEERING INDUSTRY

The designing and making of precision machine tools was an essential concomitant of the Industrial Revolution. Without such tools the manufacturing industry could not be developed; without the power provided first by steam, later by electricity, the machine tools could not be operated. Also, until the coal-mining and iron industries had become established the raw material was not available from which to make the tools and the machinery needed. All these factors combined to found and develop the machine tool industry in Britain in an incredibly short space of time. In 1770 traditional equipment known for centuries was still in use; 80 years later the industry was fully established, the modern machine tools had been designed and made under the guidance of a small group of engineers.

83 *Henry Maudslay, 1827*

It was during the seventeenth-century scientific revolution (page 11) that the new understanding in physics indicated possible ways of designing machines which were then needed

84 *The toolmaker's machine shop for threading screws. Engraving from* Diderot's Encyclopaedia, *1763*

85 *A steam hammer at work. Painting by James Nasmyth, 1871*

for the accurate measurement of time and distance. Over many years the craftsmanship evolved to produce these machines to make, for example, clocks and telescope lenses. But, though these machines were more accurate than before, the designs were still based on traditional concepts. For instance, the oldest machine tool, the lathe, the use of which dates from *c*.700 BC, was only slowly improved from the thirteenth-century pole type to a treadle machine and thence, by the mid-eighteenth century, to a screw-drive precision instrument (84).

It was in the early decades of the nineteenth century that machine tools began to be made to a sufficiently high degree of accuracy to manufacture the metal parts required by the industrial development resulting from the power of steam. One of the leaders in setting such standards of precision was the engineer Henry Maudslay (1771–1831) who began by working for that versatile inventor Joseph Bramah (1748–1814) patentee, of among other things, a new water closet (page 126), a hydraulic press and a patent lock. The fundamental invention among machine tools was that of the screw-cutting lathe, pioneered in mid-eighteenth century. Maudslay (83), who set up his own firm in 1797, in the same year produced a precision design. A few years later he made a bench micrometer with an accuracy of 0.0001 inch; this is indicative of the speed of advance in the search for precision. In 1776 the ironmaster John Wilkinson (1728–1808) was setting a new high standard by being able to bore a 50-inch diameter cylinder for the Boulton and Watt steam engine, accurate to one-sixteenth of an inch (page 30), yet Joseph Whitworth (1803–87) in 1851 was able to display at the Great Exhibition his measuring machine for which he claimed one-millionth of an inch accuracy. Like Whitworth, James Nasmyth (1808–90), an engineer of great originality, was trained in Maudslay's workshop. He is best known for his powerful steam hammer designed in 1839 which could be lowered under such complete control that, as described in the Great Exhibition Catalogue 'it would descend with power only sufficient to break an egg' (85, 86).

During the last 120 years technology has revolutionized the engineering manufacturing

86   *James Nasmyth*

industry in the development of mass and automated systems of production and in the new materials and processes introduced. Manufacture has been markedly speeded up by, firstly, the introduction in the later nineteenth century in America of the system of interchangeable parts and, secondly, the design of machines which were able to carry out more than one function, so economizing in time and labour. Henry Ford (1863–1947) introduced the assembly line system in order to mass-produce the Model T car. Each car was manufactured through all its stages in one factory instead of, as previously, the parts being brought from many factories to one centre for assembly. Since that time, mass-production systems have concentrated on reducing the time used in manual handling and moving parts.

As the speed of cutting tools in manufacture was increased a steel of tougher quality was needed. Alloy steels were produced for this containing tungsten and vanadium (page 70). In 1900 high-speed steels, containing tungsten, cobalt and chromium, were introduced which retained their hardness even at red heat, so made even faster cutting possible. The advent of tungsten carbide tools in 1926 opened up further possibilities. Such tools, made by grinding to a mixed powder such metals as titanium, vanadium, cobalt and molybdenum then subjecting it to high pressure at high temperatures, are only slightly less hard than diamonds.

A number of especially hard materials have been produced for different purposes. An early example was silicon carbide (trade name carborundum) which was intended as a very hard abrasive in substitution for diamonds. It is made by heating a mixture of sand and coke in an electric furnace. Boron carbide is even harder but much more costly. Diamonds are the hardest of all known materials and have long been used for cutting. Because of their high cost many attempts have been made to produce them synthetically. This was achieved in the USA in 1955 by means of subjecting a form of carbon (generally graphite) to an extremely high temperature (up to 3,000°C) under very high pressure (up to 100,000) atmospheres. New ceramic materials have been made to resist the high pressures, speeds and temperatures encountered in the development of rockets and nuclear power (page 39). Pyroceram is one of these, introduced into Britain from America in the 1960s (trade name Pyrosil) in the form of ovenproof, flameproof kitchenware.

Of the greatest importance in the advance of the machine-tool industry has been the replacement of steam power by electricity and consequent fitting of electric motors into much of the equipment. Electric welding has likewise represented a great advance.

THE CHEMICAL INDUSTRY

A small chemical industry has existed for centuries to supply the requirements of man in the fields of dyes, glass-making and agriculture, but it was only with the impetus caused by the Industrial Revolution and the consequent population rise that the needs of manufacturing industries became so extensive. There were two-

fold: a much larger quantity of substances which had been produced for years, alkalis and mordants for example, and a variety of different substances to supply newer industries, for instance phosphorus for matches, sulphuric acid for fertilizers, bleaching and treatment of sewage. Since the late eighteenth century the chemical industry has been built up on the application of chemical scientific theory to a means of creating industrial processes which transform a wide variety of raw or waste materials into essential products. The chemical industry above all others has had and still exerts the widest and most important influence on everyday life. In 1800 it was providing the substances urgently required in quantity by five chief industries: soap-, glass-, paper-making, textiles and agriculture. Today it still supplies these needs, but to these have been added the requirements of the manufacture of detergents, plastics, adhesives, insecticides, man-made fibres, paints and a host of others which are accepted as essential to our domestic well-being.

Manufacture of alkalis

In the later eighteenth century the expansion in the textile (page 147), paper (page 152), glass (page 79) and soap-making industries (page 77) produced a demand for alkali which could not be met. The word derives from the Arabic *al-qaliy*, meaning calcined ashes, and for centuries was used to describe substances such as soda and potash which were traditionally derived from burning certain sea-shore plants grown in a salty soil and seaweed. 'Kelp' is the name of a large brown seaweed and kelping (the production of alkali by the incineration of seaweed) was carried out on a large scale, notably on the coasts of Ireland and western Scotland. Also, in certain areas, natural soda occurred; the earliest known instance of this was on a lake-shore in the desert near Cairo, used from very early times for glass-making in Egypt.

The shortfall in eighteenth-century demand was finally met by making soda from salt (i.e. sodium chloride), two substances which are chemically related. Soda was first successfully and commercially produced in this way by the French chemist Nicolas Le Blanc (1742–1806) in

1789. In Le Blanc's process salt was treated with sulphuric acid to become sodium sulphate, then roasted with coal and limestone, and the resulting black ash treated with water to dissolve out the soda. The Le Blanc process was set up in England in the early nineteenth century in Cheshire, Lancashire and Tyneside, areas where salt deposits, also supplies of coal and limestone, were readily available. After further development by others, it was superseded by the more satisfactory ammonia-soda continuous method first successfully demonstrated by the Belgian chemist, Ernest Solvay (1833–1922), in 1861 in which a mixture of carbon dioxide and ammonia gases is introduced into a concentrated salt solution to produce sodium bicarbonate which is treated to convert it to soda.

A soft soap for commercial use was made in England from the Middle Ages. Hard soap for toilet needs was imported from Europe but was so expensive that most families made it at home. The method of making soap then and now has changed little in principle, the constituents being fats and alkalis. At home rendered-down animal fats were boiled together with lye, which was an alkaline solution made by running water through wood ashes. Commercial large-scale manufacture of soap began in the nineteenth century after soda became available from Le Blanc's process and, by the 1890s, pleasantly perfumed soap could be purchased from manufacturers such as Lever and Pears.

Sulphuric acid

Sulphuric acid is a very important industrial chemical and has been so ever since Le Blanc developed his soda process which utilized the acid in quantity. The great nineteenth-century German chemist Justus von Liebig (page 47) opined that consumption of sulphuric acid represented a true measure of a country's prosperity: this is still the case. Sulphuric acid had long been used in dyeing and bleaching among other trades; today it is fundamental to a wide range of manufacturing processes in industry from rubber (page 83), detergents (page 79 and superphosphate fertilizers (page 47) to pesticides and medicines.

As early as the fourteenth century, sulphuric acid was being made by heating green vitriol (ferrous sulphate) which gave a highly concentrated acid called oil of vitriol. This method produced only small quantities at high cost and by 1700 could not satisfy demand. Soon it was discovered that a dilute acid could be formed by burning sulphur and saltpetre in large glass jars containing a little water. This was done in England in 1736, but still demand exceeded supply. Ten years later, the British chemist, John Roebuck, (1718–94) devised the lead-chamber process in which these (which were cheap and immune to action by the acid) replaced the glass jars. By this means sulphuric acid could be produced cheaply in quantity. During the nineteenth century, the contact process superseded the lead-chamber method. In this sulphur dioxide was made to combine with oxygen in the presence of a catalyst (a substance which accelerates a chemical reaction without itself being consumed in it). In 1831 Peregrine Phillips patented this process using platinum wire as the catalyst. Technical problems took some years to resolve, but by the 1880s the process was being used commercially.

Chlorine

The element chlorine was first discovered in 1773 by the Swedish chemist C.W. Scheele (1742–86) and its usefulness as a powerful bleaching agent first appreciated by the Frenchman C.F. Berthollet (1748–1822). Before the end of the eighteenth century the use of chlorine had revolutionized the bleaching of textiles. In modern times chlorine is widely used in quantity for bleaching and to disinfect water as well as being vital to the manufacture of synthetic fibres, antifreeze fluids and polyester resins. The element fluorine, first prepared and isolated in 1886 by the French chemist Henri Moissan, has also become important industrially. Apart from its benefits in prevention of dental decay (fluoridation of drinking water), fluorocarbons, which are compounds of carbon and fluoride, are used extensively in production of some plastics and in making aerosol propellants (page 79) and refrigerants: freon (a trade name) is the best known of these compounds (see also hydrofluoride page 81).

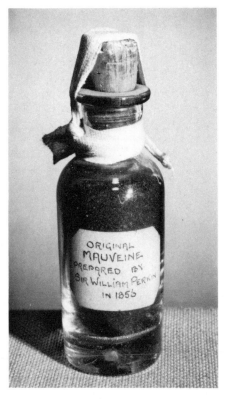

87   *Instantaneous light box, early nineteenth century. Contains chlorate matches which had to be dipped in vitriol (sulphuric acid) in order to be ignited. Bottle of vitriol in centre of box*

88   *Original mauveine prepared by Sir William Henry Perkin in 1856. The first synthetic aniline dye (page 147)*

Phosphorus

Phosphorus was first discovered in 1669 by a German alchemist Hennig Brand, who made it by heating evaporated urine with sand; then, a century later, it was found that it could be prepared from bone ash. This type of white phosphorus came into demand in the 1830s for the match industry and for many years its highly poisonous characteristics took an appalling toll of the health of match factory workers. In 1845 the Viennese Anton von Schrotter (1802–75) discovered how to convert white phosphorous into the non-toxic amorphous red form and ten years later this was used in the Swedish Johan Lundstrom's safety matches which, by dividing the chemical constituents between the match-head and the box striking surface, reduced the chances of spontaneous combustion. Today, phosphorus is also used in other manufacturing processes, notably plastics, the oil industry and in agricultural chemicals.

The chemical industry has been of the greatest importance to dyeing processes from the produc-tion of mordants in early times to the making of synthetic dyes from the nineteenth century onwards (page 147). Since the 1780s it has also contributed greatly to the manufacture of fer-tilizers for growing food. Of particular import-ance has been the production of sulphuric acid, ammonia, potash and phosphorus (page 47).

New materials

In the twentieth century, and especially since 1945, demand has led to chemical research which has opened up new fields and produced new materials which affect many aspects of daily life. Silicones are a notable example of this. Silicon is an element closely related in its properties to carbon. Silicones are synthetic polymers pro-duced by replacing carbon in these compounds by silicon. The resulting silicones possess some unusual and remarkable characteristics which are useful in a variety of ways. For example, textiles are made water-repellent by immersion in a silicon solution so stay clean longer; also silicones give fabrics crease-resistance. Silicones used in polishes give a higher gloss and retain fluidity in lubricants over a wide range of temperatures. Silicone rubbers preserve their qualities over a similar range so are widely used in ovens, washing machines and irons.

Aerosol sprays are another innovation, introduced into Britain in 1956. These pressurized containers use liquified or compressed gas as a propellant, the spray mechanism dispensing a mist (an aerosol*) of hair lacquer, insecticide, furniture polish, deodorant, lubricant or even salad dressing.

The synthetic (soapless) detergent possesses such remarkable cleansing properties because of its molecular structure. The use of these compounds, with long-chain molecules, reduces the surface tension of the water and increases the wetting power of the cleanser so, unlike soap, will produce a good lather in hard, cold or salt water. The first synthetic detergent was produced in 1916, but it was the development of the oil industry which led to widescale manufacture in the 1950s. Biological detergents, developed ten years later, contain enzymes which help to remove protein stains (egg, blood, etc) by reacting upon them.

GLASS

It is not known for certain when and where glass was first made; probably fusion of sand and soda took place accidentally on an open fire. The earliest known use of glass (*c*.4000 BC in the Middle East) seems to have been as a green glaze applied to beads and utensils to give them lustre. Glass vessels date from *c*.1500 BC and moulded glass rather later. The ingredients needed to make glass then were, as now, soda ash, lime and pure silica sand. The purity of the sand is vitally important as even traces of impurities will colour the glass or affect its consistency and transparency.

In order to make glass the raw materials have to be heated to a sufficiently high temperature to fuse them together. After the molten glass has been shaped by moulding or blowing, it must then be cooled gradually in a controlled manner to avoid crystallization and fracture; this process

is called annealing. From the Middle Ages in Europe the glass-making furnace was generally round and built of stone or brick in tiers. It was heated through the centre by burning wood so the glass-maker built his furnace in forest areas and transported his raw materials to it. The glass was melted in clay pots and, as it was difficult to achieve the high temperature, the process had to be carried out in stages. The furnace was subdivided into compartments, one for fritting, that is melting the materials into a partially-fused lump, the next for further melting and the third for annealing (90).

The vital discovery in glassmaking was the use of the blowpipe (a hollow iron tube about four to five feet long), which is believed to have come from Syria in the first century BC (89). In blowing glass, the glassmaker picks up a 'gob' of molten glass on the end of the blowpipe and blows it into a bubble. He attaches a punty (a solid iron rod) to the other side of the bubble as a handle and shapes the vessel he is making with blocks, tools and shears. (For flat glass see page 94). He also makes use of centrifugal force by twirling the pipe round in the air while the glass is still malleable (90).

In the early seventeenth century the depletion of the English timber stock became so serious that James I forbade the use of wood as fuel for various industries, of which glass-making was one. Furnaces then had to be heated with coal and, as transport was so difficult and costly, it was cheaper to move the glass-making concerns. These were set up in coal-producing areas (south Lancashire and Tyneside) and where also pure sands were locally available.

The English glass industry flourished at this time, partly from the import of foreign craftsmen and partly by the development of lead glass by George Ravenscroft. In the 1670s he was attempting to produce an English equivalent of the famous Venetian *cristallo*, a clear crystalline glass developed in the fifteenth century. He first used crushed flints (so making flint glass), but as these caused fine crazing he started to use lead oxide. This resulted in a heavy glass of great brilliance ideal for cutting and polishing.

There were a number of advances in glass-making during the eighteenth and nineteenth

---

*Aerosol is a suspension of small liquid or solid particles in a gaseous medium such as air.

89   *Glass bottle-maker's chair with blowing irons*

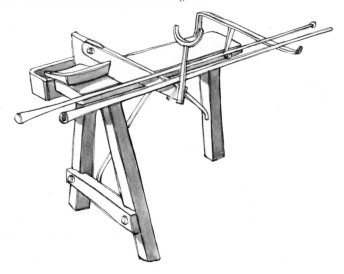

90   *Glass furnace in operation showing blowing and swinging and flattening the parison (the blob of molten glass). From* Agricola, 1556. *Note: A blowing irons, D forceps and E mould.*

91   *Blowing glass goblets as shown in* Diderot's Encyclopaedia, 1763. *On the left the blower is spreading the roll to make the goblet base and on the right he presses the base onto his apprentice's punty in order to disconnect his own blowpipe*

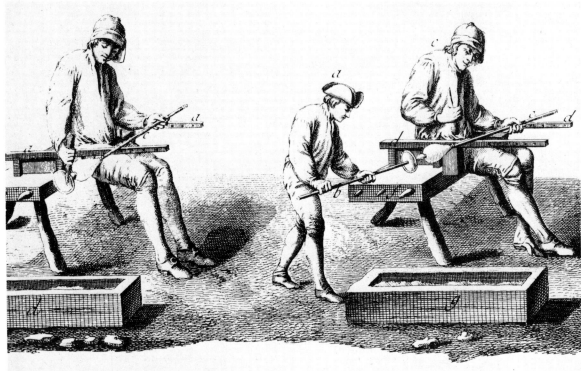

centuries. In England the cone-shaped glass-house was introduced which was more efficient in its use of fuel as the air currents were drawn up into the higher structure. Research into the chemistry of metals and of colouring agents led to the introduction of new and richer colouring of glass. For example, rich ruby shades were produced from an alloy of copper, zinc and manganese oxide, a bright yellow was derived from uranium and a brilliant green from chromium. Nickel was used to give various soft shades from brown, through grey, to purple.

Decoration of glass

There are many ways of decorating glass artefacts. Some are traditional, the craft dating back over centuries, others have arisen as a result of a new technology. Among the former cutting* and engraving have been practised since the sixteenth and seventeenth centuries. The decoration was marked on the vessel (92), then it was cut by an iron wheel fed with sand; in engraving, smaller copper wheels were employed. Afterwards the vessel had to be polished. Later techniques utilized carborundum and, more recently, engraving wheels are impregnated with diamonds (94). Since 1970 the process may be automated with computer control (page 168). Other twentieth-century decorative methods include sandblasting, carried out inside a cabinet, where the surface of the glass is attacked by blowing an abrasive at it through a nozzle, printing a picture by silk-screen process (the first automatic machine for this was introduced in 1950) and etching with the use of acids (generally hydrofluoric), a wax resist being applied to the area not to be treated. A do-it-yourself (DIY) glass etching kit appeared on the market in 1980, so plain glass vessels may be decorated at home. Solution (guaranteed harmless to the user), resist and stencils are supplied.

By about 1650 the glass bottle began to replace stoneware for ale and wine. In the nineteenth century mineral water was sold in sealed glass bottles and in the 1880s the milk bottle appeared. These bottles were blown, each blower producing about 3,500 a day. Efforts were made to mech-

*The great age of English cut glass was 1750–1810.

anize the process; the Ashley bottle machine appeared in 1886, followed in 1903 by Owens' automatic version. Today the whole proceeding, from molten glass to finished bottle, is mechanized.

Modern glass manufacture

Twentieth-century chemical research has shown how to make glass with different properties suited to a variety of needs. Borosilicate glass (containing borax), which resists sudden changes in temperature, was developed both for the laboratory and the kitchen. A whole new glass-making process was needed to meet the demands of the television industry (page 164). For example, cathode ray tubes are produced automatically with glass of even thickness to withstand pressure, of special composition to prevent the passage of X-rays generated by the incidence of the electron beam through the tube and are sealed electrically. Vacuum science also has made particular demands. In the home this applies especially in the making of electric light bulbs and vacuum flasks. For the bulbs a transparent envelope is needed which must be vacuum tight to prevent oxygen and water vapour from the atmosphere reacting with the hot filament.

The development of glass fibre is a remarkable example of modern glass technology. It has long been possible to make coarse threads of such fibres, but their commercial production into very fine filaments dates from the 1930s. Today there are two main types, the continuous filament which is widely used as a reinforcement material, especially in plastics (page 85) and glass wool which is ideal for insulation of, for example, roofs and pipes (page 107). For making the continuous filaments high quality low alkali 'E' glass is used; it is melted at a high temperature (c. 1,250°C) then drawn mechanically through bushings at some 12,000 feet per minute, which can produce threads as fine as 0.00003 inch diameter (93) (page 85). Soda borosilicate glass is used for wool fibre; the molten glass, rotated at speed in a dish, is thrown out by centrifugal action into the tiny holes of a spinner to emerge as short fibres. These are formed into a mat for insulation by the application of a binding agent (128).

92   *Marking out a glass goblet to prepare it for hand-cutting*

93   *Drawing fibreglass continuous filament for reinforcement purposes*

94   *Deep cutting by holding the jug on an abrasive wheel*

RUBBER AND PLASTICS

Rubber

Natural rubber is a latex found in the inner bark of several different plants and trees the chief of which grew wild in South America but, from the 1880s, were cultivated in plantations, mainly in Malaya, Indonesia and Ceylon. This latex was used by the South American Indians many centuries ago to make rubber balls and one-piece footwear; certainly the Spanish *conquistadores* were familiar with it. In Europe it remained a curiosity. In 1770 Joseph Priestley recommended its use as a pencil eraser, so in England it became known as 'india rubber'.

Serious interest in the properties of rubber and its possible commercial uses were only aroused by the scientific study expeditions to South America undertaken by the French, the results of which were published in the 1750s. The problem in preparing natural latex (called by the French *caoutchouc* from the local Indian term meaning 'weeping wood') for useful purposes was that it quickly hardened on exposure to air and so, when imported, was a solid substance which lacked malleability and reacted badly to a range of temperatures. Many attempts were made between 1750 and 1820 to find a suitable solvent for this solid rubber which would make it easier to handle in manufacture. Some success was attained and small factories established to incorporate rubber into braces and garters, but it was Thomas Hancock (1786–1865) whose experiments and inventions in England helped to found a rubber manufacturing industry. In 1820 he opened up a shop in London to work and cut rubber for use in garments, and he invented the masticator, a machine which softened the raw rubber by kneading, so making it malleable for handling. He moulded rubber under heat and pressure and used liquid latex to saturate fabrics for waterproofing.

Meanwhile the Scottish chemist Charles Macintosh (1766–1843) produced a waterproof fabric which he made by cementing two thicknesses together with rubber dissolved in naphtha. Macintosh patented his process in 1823 and the following year opened a factory in Manchester to produce 'mackintoshes'. In the early 1830s rubber production was started up in America and from here came the important discovery by Charles Goodyear (1800–60) of vulcanization. In this process (patented 1844) he treated the rubber at high temperature with sulphur and so increased its elasticity and strength, which it retained over a wide range of temperatures. Before long rubber was being produced in sheets and extruded as tubing and thread.

During the later nineteenth century rubber was used chiefly for road tyres, cable insulation, footwear and waterproofing. The application of chemical science in the twentieth century led to improved quality and an extended range of rubber due to the introduction of various compounds into the material, for example, foam rubber appeared in 1928. In the 1930s latex emulsion began to be shipped in tankers direct from the plantations, making a much wider use possible in elasticated corsetry and underwear. By the late 1950s a number of excellent synthetic rubbers became available such as silicone rubbers (page 78) and elastomers*, for example, Lycra an elastic fibre produced by Du Pont in America. This was quickly manufactured into a varied range of underwear, stretch garments and furnishing fabrics.

Plastics

No material has ever changed the appearance of the home as much as plastics have done and not only their appearance but also their resistance to wear and easy-care characteristics have also altered everyone's way of life. The word, which derives from the Greek *plastikos* meaning 'that which may be moulded', describes their quality of being easily shaped to any form desired. Today, most plastics are manufactured from chemicals derived from oil or coal, but the plastics story began in the nineteenth century when they were produced from natural materials.

All plastics are polymers. This means that, whether natural or man-made, their structure resembles a chain consisting of large molecules each of which is formed from smaller molecules. The word polymer comes from the Greek *polus*

---

*A class of synthetic polymers possessing elastic qualities.

95   *Kitchenware made from 'Propathene', ICI's trade name for polypropylene*

96   *Garden equipment made from 'Alkathene', ICI's trade name for polyethylene*

meaning 'many' and *meros* meaning 'part'. Natural polymers were used to make the early plastics, materials such as cellulose or milk, but modern plastics are synthetic, created by a process of polymerization in which the small molecules are chemically combined artificially into chains of larger ones. Many of the names of modern plastics are prefixed by 'poly', indicating which type of molecules they have been polymerized from, for example polystyrene from styrene molecules and polyvinylchloride (PVC) from vinyl chloride molecules.

The first plastics were derived from cellulose nitrate made from paper, wood or cotton fibre dissolved in nitric and sulphuric acids. Parkesine, produced by Alexander Parkes and shown at the Great Exhibition of 1862 in London, was made into decorative combs, buttons and knife handles. Celluloid (used particularly for billiard balls and collars) followed in 1870 and casein, derived from milk and rennet cured with formaldehyde, in 1900. In the twentieth century both cellulose nitrate and acetate were being employed for photographic film and in the 1920s the fibre form was developed as acetate rayon (page 148). The first completely synthetic plastic was Bakelite. This was the trade name for a phenol resin patented by Dr Baekeland in America in 1907. In its characteristic dark mottled browns and greens Bakelite was made into a range of electrical and household goods in the 1920s and 1930s; especially typical were cabinets for wireless sets and clocks (page 160).

Although much of the early work on modern plastics was done in the 1930s it is only since 1945 that this has been speeded up and the materials produced in quantity. High-speed, mass-production methods are now employed, the material being softened by heat, then forced into metal moulds of the required shape and allowed to harden. Various means are adopted to do this, for instance it may be extruded, spun, sprayed, rolled, injected or compressed. The three types of plastic most often used for domestic needs are polyvinylchloride (PVC), polyethylene and polystyrene. These possess differing characteristics suitable for various purposes. PVC (often called just vinyl) appears in a wide range of products because it can be made in so many forms, either rigid or flexible, thin or thick, transparent or opaque. The rigid form appears in pipes and guttering, also gramophone discs, while a more flexible version is made into floor tiles or soles for footwear; if softer still it can be used for 'wet look' garments, handbags, wallpaper, hosepipes, furnishing fabrics; lastly, as the ubiquitous transparent film it is useful for wrapping food in the shops and at home.

Polyethylene (polythene) comes in two main forms, the low density version first produced in the 1930s, which is used in sheets to protect and wrap almost anything, and the high density type, developed in the 1950s, which is heavier and more rigid; this is invaluable for making bottles, washing up bowls, buckets, dustbins (96). There are three chief types of polystyrene: the toughened version seen particularly in refrigerator linings; the clear type which is an important packaging material seen in egg cartons, yoghurt and butter containers and a variety of pots and boxes; and, thirdly, the expanded polystyrene, a lightweight foam substance which is moulded so that it packs and protects any article, however delicate.

Apart from these three plastics present in everyone's home, there are others too in common use: polypropylene, a rigid thermoplastic with excellent resistance to liquids and solvents, which is inexpensive and suitable for kitchenware (95); phenolic resins still employed for electrical and other fittings (polyurethane, page 149); melamine resins for unbreakable crockery; and many plastics, such as nylon and polyester, made into clothing (page 148). Polyester is also the plastic most often strengthened with glass fibre (page 81). This is a tough new material known as glass-reinforced polyester which is ideal for a variety of purposes from piping, water tanks and roofing to trays and lampshades. Of tremendous help in cooking has been the development of polytetrafluoroethylene (PTFE), which is inert to a wide range of chemicals and resistant to sunlight and moisture. It is used to give a non-stick coating to the inside of cookware.

CERAMICS

In modern parlance this term covers a diverse group of products which includes glass, cements and plasters, but traditionally, and certainly domestically, it comprises articles made from clay of a plastic consistency, dried and fired for durability. Earthenware (pottery) vessels and utensils have been made since early Neolithic times for cooking and storage. Earthenware clays are common but, as they contain many impurities, can only be fired at comparatively low temperatures (up to 1,000°C) to produce a soft pottery.

In Europe only this softer earthenware was manufactured until knowledge of how to make harder ware was acquired during the Renaissance. Articles were formed by hand, either by pressing the worked clay into a mould or by coiling it, that is building up the shape by curving ropes of clay round upon each other then smoothing the finished article. The great breakthrough came with the potters' wheel which originated in the Middle East about 3000 BC. With this technique, set in motion by the feet, vessels of a more sophisticated form could be produced much more quickly (97).

All clay artefacts need to be dried before firing. In primitive communities they were baked in shallow pits, but by 5000 BC the kiln had been devised to fire at a higher temperature in order to make a more permanent article. Kilns are furnaces comprising a firebox in which wood, and later coal, was burned, a chamber where the pots were set out and a flue for escape of gases.

Earthenware was decorated by designs incised in the plastic clay, by relief modelling added to the surface and, most commonly, by slip. This is clay mixed with water to produce a creamy liquid either in the same colour as the vessel or different.

Earthenware was made impermeable to liquids by coating it with a glaze. The origin of glazing is very ancient: Egyptian glazes were made of clear or coloured glass fused on to the pot in a second firing (page 79). Later a variety of chemicals – iron, manganese, copper, cobalt – were incorporated in the glaze to give improved practical and artistic colour effects. Lead glazes, made from a fusion of pure silica (from glass) and lead oxide were employed from early times. It was found (in the Middle and Near East) that the addition of tin oxide rendered the glaze white and opaque,

virtually an enamel, and so produced a whiteware which could be decorated attractively in colour. Knowledge of this tin glazing was brought to Europe by the Moorish invaders and was experimented with in the thirteenth century in Spain. Such ware from the island of Majorca was imported into Italy and Italian potters were soon making 'maiolica' ware which was later exported to England. One Italian town where the ware was made was Faenza. Potters from here settled in the sixteenth century in France, Spain and the Low Countries; their pottery was called faïence. By the mid-seventeenth century the tin glaze industry was centred on Delft in Holland whose more delicate ware, beautifully painted, often in blue, became famous. A similar ware was being produced in England at this time.

Meanwhile efforts had been made since the Middle Ages to produce a harder, more durable article. This was achieved in the sixteenth century in the German Rhineland and was widely exported. This stoneware (then called Cologne ware) was made from a purer clay, containing more silica and some felspar. It was fired at a higher temperature (1,300°C) and was salt glazed, the salt being thrown into the kiln during firing. Such glaze is hard and possesses a high gloss. In England stoneware was first made in 1671 at the Fulham Pottery in London by John Dwight and from 1680 onwards became a staple Staffordshire product (99). But stoneware was only a stepping stone on the way to the goal of European potters which was to produce porcelain of the quality of the Chinese material. Stoneware was ideal for the average home, but the more costly porcelain was hard, yet delicate and beautiful.

Porcelain is made from a compound of kaolin, ball clay, felspar and silica. The hard paste (true) porcelain is fired at a high temperature (1,400°C). It was made in China as early as the ninth century, and imported ware had long excited envy and admiration in Europe when in 1710, the German chemist Johann F. Böttger (1682–1719), who worked at the Meissen factory near Dresden, discovered a method of making it. He added ground alabaster or marble to the local white clay. Several French scientists also worked on the problem, among them Réamur, Hellot and Macquer, and success was finally achieved at the

97   *Potter using a kick-wheel to throw pots. From* Chemical Technology *by F. Knapp, 1848*

98   *Earthenware saggars stacked in a kiln ready for firing the contents*

100   *Painting porcelain, 1771 from* L'Art de Porcelaine *by M. le Compte de Milly, Paris*

99   *Salt-glazed stoneware loving cup, English, 1740*

State Manufactory at Sèvres in 1768 (100). Meanwhile in England the chemist, William Cookworthy (1705–80), was also successful, after experimenting since 1745 with Devon and Cornish clays.

An alternative line of research had been pursued in the making of soft paste (*pâte-tendre*) or artificial porcelain. This generally contains frit (a powdered glass) in substitution for felspar and is fired at a lower temperature (c.1,100°C). An early version had been produced in Italy by the Medici in Florence about 1585 from kaolinic clay and the French developed it using frit in the seventeenth century. Production began in England in the 1740s at Stratford-le-Bow with a kaolin imported from America and before long soft-paste porcelain was being made at Chelsea and a number of other cities. It was at Chelsea that bone-ash was incorporated into the mixture to produce a harder, more translucent ware.

By the early eighteenth century, as timber supplies failed, the pottery industry experienced the same fuel problems as other industries had done, glass and iron for example (pages 67, 79). Coal, as the replacement fuel, was costly and difficult to transport before the age of railways and availability of a good road system, so, following the example of other industries, the potteries moved to where fuel and raw materials were available together, in this case Staffordshire. Earthenware had been made in this area on a small scale for a long time and stoneware since 1680. Gradually, empirically, this was made into a more desirable, hard, light-coloured ware which became a substitute for imported porcelain or Delftware, so making Staffordshire an important ceramic centre.

The first step forward was the use, from about 1710, of a white-burning clay from the West Country which, mixed with fine grit and sand, gave a light-coloured stoneware. Then, in 1720, calcined flint was added, making the ware whiter still, and rendering the body more refractory so that it could be fired at a still higher temperature to make it harder. Both these advances are attributed to John Astbury (c.1688–1743) and both materials were transported to the potteries by river. The use of coal in the kilns to replace wood raised the problem of the deleterious effect

of the emission of combustion products upon the fired ware; this was solved by enclosing the pots in fireproof clay cases called saggars (98). The use of calcined flint presented a different problem, that of lung damage to potters caused by the silica dust from the grinding process; this was solved by a mill (patented 1726) in which the flints were ground under water. Then in 1750 Enoch Booth introduced double firing, which gradually replaced salt glazing. In this process the pottery was first fired to the biscuit stage; it was then decorated and glazed and fired again. This proved a more reliable method than the addition of salt had been.

In the second half of the century further technical improvements were made in the composition of the clay body, in its drying, in kiln design, in control of temperatures, in new means of decoration. For example, transfer printing was in use by the 1750s. In this the pattern was transferred from an oiled copper plate to the pottery article via a thin paper print. Also, about 1745 pot throwing was supplemented (for cheaper, mass-production standardized ware) by the use of moulds, first made of metal, later of plaster taken from an alabaster master. The industry expanded greatly as scientific methods gradually replaced empirical ones.

The great potter, Josiah Wedgwood (1730–95), contributed most at this time to solving the problems of pottery manufacture by technological and scientific means (102). Entrepreneur, classicist and experimenter, Wedgwood built his new factory 'Etruria' in 1769 and ran it on advanced lines for fast, efficient, cheap production of well-designed quality ware (101). In 1762 he had produced his improved stoneware, (with the permission of Queen Charlotte called Queen's Ware) (103) and went on to develop his coloured jasperware ornamented with classical reliefs in white, chiefly designed by the sculptor John Flaxman. These wares are still manufactured today. He introduced steam power in 1782 to mix clay, grind flint and turn lathes.

Wedgwood had brought elegance to many a breakfast table at a fraction of the cost of imported porcelain. His contemporary Josiah Spode (1733–97) also produced a cream pottery (he specialized in a blue printed ware) which

101 *The Etruria factory built for Josiah Wedgwood by Joseph Pickford 1768–9. Opened 13 June, 1769*

102 *Josiah Wedgwood, 1780. Portrait in enamels on Queen's Ware by George Stubbs RA*

103 *Cream-coloured Queen's Ware tureen. Wedgwood, late eighteenth century*

could be sold cheaply to everyone. His son, Josiah Spode II (1754–1827) improved this in 1797, making a hard-paste ware with the addition of bone-ash and felspar, so establishing the bone china for which England is famed. This inexpensive, attractive and durable ware was in great demand with the population explosion of the nineteenth century. At this time also ceramics were needed in even greater quantity as new industries developed with new requirements. From the 1860s onwards, sanitary ware was needed in quantity for kitchens, bathrooms and lavatories (page 124), the chemical industry needed large containers, the electrical and tele-graphic and steel industries made new demands.

To meet these requirements the pottery industry was gradually revolutionized. The trend towards scientific research and technical innovation, started by Wedgwood, snowballed. More processes were powered, first by steam, then gas and electricity. The design of kilns was changed, temperatures were more carefully controlled, raw materials and chemicals for colouring and glazing were scrutinized and analysed to improve performance, new decorative printing methods were evolved and the pottery worker was better protected against the dangers of chemicals, such as lead, used in glazes.

CHAPTER FIVE

The Home 2 Exterior and Structure

TRADITIONAL BUILDING MATERIALS FOR HOUSING BEFORE 1800

Apart from medieval fortified castles and manor houses, and also, later, palaces and great houses for the wealthy landowner, most homes were constructed from local materials. In areas where stone was plentiful and easy to quarry it was used for housing, and many beautiful examples survive, but stone was costly and difficult to transport far. In chalk regions houses were often faced with flints, so increasing their durability as well as lending a pleasing form of decoration. During the Middle Ages timber was used everywhere, often with plaster or brick infilling. Wattle-and-daub sufficed for poorer dwellings. Then, as timber supplies began to diminish, brickwork skills were developed and the material was widely used where stone was not readily available. Roofing was by thatch, timber or ceramic or stone tiling.

Wattle-and-daub, unburnt earths, thatch

The use of these simply made, cheap materials was widespread from very early times. Wattle-and-daub provided a primitive means of walling made from a row of vertical stakes or branches (wattles) interwoven horizontally by smaller branches or reeds. Clay mud (daub) was plastered on one or both sides and dried in the sun; it was then reinforced with turves and moss. This structural method was used throughout the Middle Ages, the branches being later replaced by laths, with hair and straw being incorporated

into the mud to give strength and durability. It was long employed as infilling to timber structures.

Several ways of using clay earths for building were evolved over the centuries. Some were malleable substances made by adding water and further materials; in others the earth was rammed or beaten against a hard surface until it cohered. Different names were given to the pliable forms (for example, cob in south-west England) as the 'mixes' varied from district to district. Lime had to be incorporated into the mix to make it set; straw and sand also added to the cohesion. For strength an aggregate was added, gravel or fragments of slate or stone. Sun-dried 'bricks' were also made of clay earth mixed with water and straw, then pressed into wooden moulds and air dried. These large 'bricks', about $18 \times 18 \times 6$ inches, were used for walling bound together by a clay mortar.

Buildings made of these materials were based upon a low plinth of stone, brick or rubble to keep the base dry and prevent invasion by vermin. Such structures would then last for centuries. Roofing was by thatch, its light weight being suited to walls which would not bear a great load. From earliest times buildings had been roofed by turves, moss, heather or reeds; by the Middle Ages reed or straw was in general use. The chief problem was the danger of fire and because of it, restrictions on thatching in towns came into force as early as the thirteenth century. In rural areas such roofing material has survived

105   *Timber-framed window, pargetted panels. Port Sunlight Estate, Wirral*

104   *Oak timber-framed house, late fifteenth century. Wattle and daub infilling (daub of mud, manure and chopped straw). Re-erected 1967 in Avoncroft Museum of Buildings, Worcestershire*

106   *Flint and stone chequerwork. Kings Lynn, fifteenth century*

107   *Terracotta panel, 1875*

108   *Stone house, Athelhampton, Dorset, from c.1485*

till the present day, the thatcher's craft being developed over centuries to an advanced and decorative standard (109).

109   *Thatching. Beating up the straw*

Timber framing

Wood is one of the oldest and most-used building materials. In England, as late as the seventeenth century, the majority of houses were constructed wholly or partly from wood and even in towns, where fire was an ever-present hazard, this was still so. During the Middle Ages, forests still covered an immense area of the country, nearly all of the timber being hardwood and much of it oak, which was regarded as the best wood for timber framing.

Very early buildings were probably, like the surviving tenth-century walls of the little church at Greensted in Essex, constructed from solid half tree trunks in the Scandinavian manner but, from medieval times onwards, wooden buildings were usually half-timbered. This means that the timbers were split (halved) into square posts. The massive ground-floor storey posts were set in a

sill which rested upon a low brick or stone plinth. Upper storeys projected outwards beyond the lower one. According to district the exterior surface could be weather-boarded or the panels between posts filled with wattle-and-daub or brickwork (104). The wattle was faced with plaster, sometimes decoratively patterned with pargetting: a skilled craft especially practised in the seventeenth century (105).

Stone and flint

Until the later eighteenth century when the canal system was developed to transport heavy materials (page 96), builders used only stone available locally. There were two such large areas: firstly, the limestone regions of the Cotswolds and the Midlands, which included the famous quarries at Portland developed notably by Wren, and also the quarries of Bath which supplied so much material for the city and surrounding district. The second area comprised the Pennine regions of Yorkshire, Lancashire and Derbyshire and yielded sandstone, millstone grit and some limestone. The stone was worked into finished blocks (ashlar) for building (108) and was widely used in rubble work; it was also cut into large tiles for roofing. Granite from the west country and eastern Scotland had only a limited use for housing as it was so hard and difficult to work before, much later, power tools were introduced. Welsh and Cumbrian slate was employed extensively as a building material, in block form for the structures and split into thin slabs for roofing, paving and plumbing. More recently, it has been revived as a (fairly costly) cladding material.

Flint is an immensely durable substance which has been utilized in building since before the Romans came to Britain. It is a form of silica which is very hard but easy to split. Flints occur in chalk deposits and have frequently been incorporated into walling in large areas of southern England, especially East Anglia, where skills were perfected to ornament structures as well as making them more durable. The flints were embedded in mortar between lacing courses of stone or brick inserted to lend strength. The introduction of split flints, from about 1300 onwards, made such decorative qualities possible as whole flints are not of particular ornamental interest. The process of splitting, then trimming into the desired shape and thickness (knapping) was a skilled technique. The knapped flints were set into the mortar, face outwards, flush with the wall surface in decorative designs (flushwork) (106).

Bricks, tiles, terracotta

Brick is a convenient building material, in use for over 6,000 years; it is cheap to make, the raw materials are plentiful and it can be produced in a variety of shapes, sizes and colours. The Romans built extensively in brick, but after the collapse of the Empire the craft died out in Britain only reappearing slowly during the Middle Ages. Under the Tudors brickwork was revived and a skilful handling of the material in structure and decoration was achieved. By the early seventeenth century the demands of classical architectural forms were being met by brick craftsmen, the size of the brick had been standardized and, from Holland, came Flemish Bond design and gauged brickwork, with its fine joints and rubbed bricks. The century from 1660 to 1760 was a great age in English brick building.

Bricks are made from a variety of clays which have been puddled. In this process, after removing pebbles and grit, the clay is mixed with water and sand, then worked to give an even consistency. It is then moulded to the required form, dried and burnt, in early times in clamps, later in kilns (page 86). Puddling was at first done by men treading the material barefoot. Later the pug-mill was designed in which blades churned up the mixture in a cylinder.

Tiles for roofing were similarly made and terracotta* was introduced from Italy in the sixteenth century for moulded decorative work. This was a harder, less porous substance than brick, its mix including grog, that is fired earthenware ground to a powder. The eighteenth-century Coade Stone was similar; it contained also various fluxes and a high proportion of grog so producing a particularly hard, durable material suitable for fine decorative work. For this purpose it was employed by many

* An Italian word meaning cooked earth.

leading architects, notably Adam, Soane and Wyatt.

Glass

Windows were glazed in many Romano-British houses, but, after the departure of the Romans, domestic window glass became a rare luxury enjoyed only by the wealthy until well into the fifteenth century. All kinds of substitutes were used: mica, parchment dipped in gum arabic, oiled linen or thin sheets of horn or alabaster. By the sixteenth century domestic window glass was more often seen but was still costly.

Until the later eighteenth century flat glass was made by blowing (page 79). There were two processes, the cylinder (also called muff or broad glass) method and that of crown glass. Cylinder glass was blown into a sphere then the craftsmen would swing and twist his blowing iron in the air until the glass formed a sausage shape. A hot iron was used to cut off the ends and slit open the resulting cylinder (120). It was then cooled and flattened into a panel in an annealing oven. In crown glass the craftsman transferred the blown sphere to a punty which he then spun so that the action of centrifugal force opened up the glass into a flat circular plate which might be as much as five feet in diameter. This plate, known as a table of crown glass, was then annealed and cut up as required (110).

THE EFFECTS OF THE INDUSTRIAL REVOLUTION 1780–1900

The towns

The steady migration of agricultural workers from country to town in search of work as the Industrial Revolution got under way in the later years of the eighteenth century became a flood by the 1830s (page 14, 19). The face of this urbanization showed great contrasts. Though conditions of living for the poor were extremely bad, the development of civic planning and housing for the well-to-do was extending rapidly. Streets were laid out in terraces and squares and acres of land were covered by many fine, solid, graceful structures. Portland stone was used for the larger buildings and imitative stucco-covered brick for the less costly ones (111). At the opposite end of the social scale,

110   *Finished 'table' of crown glass*

speculators over-developed the sites which they had bought, crowding together as many dwellings as possible. The term jerry-building, the origin of which is not certain, was used to describe such sub-standard housing as early as 1869.

Not all speculative building was of inferior quality. Thomas Cubitt (1788–1855) carried out such work on a grand scale in London from the 1820s onwards, employing good craftsmen, buying good materials and erecting buildings to a high standard which nowadays are accepted as fine urban architecture. But between Cubitt's houses for the wealthy and back-to-back housing for the poor a Victorian suburbia was gradually created. These rows of terrace houses on the outskirts of cities were not by and large beautiful but they were soundly constructed and contained well-designed and -proportioned rooms. Many still stand and provide good homes.

In the second half of the nineteenth century came the beginnings of better housing for working people, schemes initiated by philanthropists, industrialists and architects of social purpose. In 1854 in Saltaire in Yorkshire Sir Titus Salt, the Bradford mill-owner, began the building of a new mill and around it housing for his workers, and also a hospital, a church, a library and an institute: the first 'new town' in Britain (113). Other industrialists followed Salt's example: Lever at Port Sunlight (105, 114) (Liverpool), Cadbury at Bourneville (Birmingham) and Rowntree near York. The experimental housing estate at Bedford Park (London) was initiated by Carr in 1876 (112) and in the 1860s George Peabody and Sydney Waterlow were providing inexpensive accommodation in large tenement blocks which, at the time, were an immense improvement on that available for working class families in inner city areas.

As late as 1830–40 the provision of sewage disposal and of water supply in the rapidly expanding towns was totally inadequate. The periodic outbreaks of cholera in London between 1832 and 1854 were largely due to the contaminated rivers which acted as open sewers and it was only after the Chadwick Report of 1842, leading eventually to the setting up of a Public Health administration and the acceptance that cholera was a waterborne disease, that a proper sewage disposal system began to be constructed when many of the old hollowed tree trunk and brick sewers were replaced by iron and glazed ceramic piping.

While the development of the engineering, the ceramic and the iron and steel industries had made the construction of such systems possible, steam power for pumping purposes was facilitating the provision of an adequate water supply. In the years 1840–50 this was still intermittent and as, later in the century more households installed baths and water closets (page 126) the strain on the system became overwhelming. Gradually dams were built, reservoirs formed and, resulting from bacteriological and chemical research, filtering systems were incorporated and chemicals added to supply all households with constant, pure water.

Structure and materials

The nineteenth century was a time of feverish building activity especially in housing, partly to accommodate the greatly increased population and partly to re-house those who migrated from country to town. Traditional building materials (page 91) were still employed but, in order to speed up construction, a tentative use was made of mass-production methods. The development of the railway and canal systems made transport of materials easier and quicker, but the tendency still was to use stone only where it was locally available or for important civic and commercial architecture in city centres and private mansions for the wealthy. As timber resources were further depleted there was a great increase in brick building. Clay was inexpensive and available in quantity. Kiln design had improved, coal was also ready to hand for firing and new techniques in brick-making were devised, such as wire-cutting (1841) and mechanical extrusion (1875). Also, tiles were being more widely used for roofing. Particularly in towns, where it was thought desirable to imitate the costlier stone, stucco facing covered the brick structure and this was then painted in stone colour. John Nash (1752–1835) was one of the architects who helped to popularize this method by his Regent's Park terrace architecture in London.

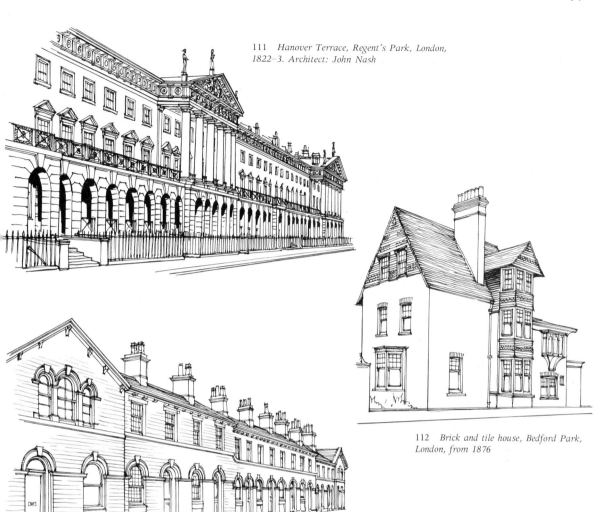

111 Hanover Terrace, Regent's Park, London,
1822–3. Architect: John Nash

112 Brick and tile house, Bedford Park,
London, from 1876

113 Stone terrace houses. Saltaire, Yorkshire, 1854–70

114 Brick, stone and tile terrace houses. Port Sunlight Estate,
Wirral, Cheshire. Founded 1888

Many architects and builders were using Coade stone to decorate their stucco-faced brickwork in imitation of carved stone (page 94). Others ornamented their unfaced buildings with coloured terracotta as in Alfred Waterhouse's Natural History Museum in London (1873–9) and several of Norman Shaw's houses (107). Doulton's of Lambeth were producing then some excellent decorative panels in this material.

Several improvements had been made in the quality and content of plaster and stucco for interior and exterior facings. Nash, for example, used an exterior stucco made from sand, brickdust, powdered limestone and lead oxide. There had also been considerable advances in the quality of mortar needed for bricklaying due partly to John Smeaton's (1724–92) experiments to find a hydraulic cement suitable to bond the stone blocks of his new Eddystone lighthouse in 1756 (115). Smeaton used a mixture not dissimilar from that the Romans had utilized for their concrete vaults which had included *pozzolana** (page 106). Smeaton's mortar was made from

115   *John Smeaton (Eddystone lighthouse on the left)*

*A volcanic ash which contains alumina and silica, and, combined with lime, produces a durable concrete. Named after its town of origin, Pozzuoli near Naples.

Welsh limestone (containing a quantity of clay) mixed with imported Italian *pozzolana*. Later, improved cements were patented, notably Charles Johnson's (1811–1911) Portland cement of 1845 in which the raw materials had been burnt at a high temperature till almost vitrified, then the resulting clinker finely ground. The invention of the concrete mixer in 1857 was of great benefit in both brick building and the increasing use of concrete as a material.

Towards the end of the eighteenth century wrought iron was being employed increasingly as structural reinforcement to timber and stonework, while the usefulness of cast iron as a moulded decorative material was being exploited in the early decades of the nineteenth (116, 117, 118). But it was from 1850 onwards that iron, both as a structural and ornamental material, came into its own. The development of the iron and steel industry was making both cast and wrought iron a stronger and cheaper material to use for a wide variety of purposes (page 66) and the erection of Sir Joseph Paxton's 'Crystal Palace' (so named by Punch) in Hyde Park to house the Great Exhibition of 1851 lent great impetus towards the extending use of this material. The Crystal Palace also demonstrated the suitability of the ferro-vitreous structure for prefabricated systems of building (119).

Though utilized less in domestic architecture than in industrial, commercial and civic building, iron was adapted for a wide range of purposes in housing which included window frames, supporting columns and beams, railings and staircases as in, for example, the early instance (1815) in the Royal Pavilion at Brighton. Later in the century every home was fitted with cast-iron fireplaces and kitchen ranges and many also boasted a selection of iron furniture and garden equipment. Because of its unfortunate predisposition to rust, ironwork had to be painted or black-leaded in order to preserve its surface, but so treated, it was almost indestructible (64, 65, 66, 67, 68, 71, 72).

Flat glass continued to be made by the cylinder and crown methods and it was not until demand for high quality glass and larger windows rose high enough to make the necessary capital outlay worthwhile that cast plate glass began to be made

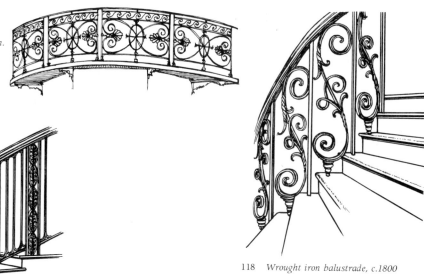

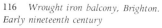

116 *Wrought iron balcony, Brighton. Early nineteenth century*

117 *Cast and wrought iron balustrade, 1805*

118 *Wrought iron balustrade, c.1800*

119 *Engraving by G. Brounger of the interior of the Great Exhibition Building of 1851 (The Crystal Palace) by Joseph Paxton. A structure of iron, glass and wood (much of it prefabricated) comprising 3,300 columns, 2,150 girders and 900,000 square feet of glass*

in Britain. Plate glass (so-called because of its earlier use for mirrors which were originally called looking glass plates) had been manufactured in England since the early seventeenth century in the same way as cylinder glass, but it was blown to a greater thickness so that it could afterwards be ground and polished to give a lustrous finish. The cost of such production was high so it remained a luxury article, too costly for window glazing.

Then the French evolved a method of casting plate glass instead of blowing it into cylinders which had to be cut and flattened out. The molten glass was run directly on to a flat table where it was rolled out. By contact with the table surface and that of the roller both sides of the glass lost their transparency and a great deal of grinding and polishing was needed to restore the lustre. The process was still lengthy and costly (121). It was in 1776 that British plate glass was made by this method in a huge new casting hall at Ravenshead in St Helens and during the nineteenth century a number of other factories were established to manufacture it. Gradually improvements were introduced, including better equipment for grinding and polishing and mechanization of the handling processes.

A method of making an alternative, cheaper flat glass was developed in Germany which was an improved type of cylinder glass. Called sheet glass, it derived from a much larger cylinder than before (about six feet long). It was ideally suited to window glazing, but implementation of the method was retarded in Britain because of the effects of the window tax which had been imposed in 1696. This tax, and the excise duties which were in force between 1745 and 1845, related to the number of window openings in houses and to the weight and quantity of glass used. This mitigated against sheet glass because it was thicker than crown glass. It was only after the tax and duties were reduced and later lifted that manufacture of sheet glass in Britain began to replace that of crown glass. In 1832 Chance Brothers of Birmingham took up the manufacture of sheet glass, cutting the cylinders when the glass had cooled, then flattening them after reheating to soften the material. In 1851 the firm supplied the 900,000 square feet of sheet glass

needed in the Crystal Palace (119, 120).

THE TWENTIETH CENTURY

Private and municipal housing

Differing means were adopted between 1900 and 1939 to house the still rapidly growing urban population in a manner which was a social improvement upon the terrible conditions of the previous century. Ebenezer Howard propounded the idea of the garden city in which a town should be built on land owned or held in trust for its community. An essential feature of Howard's vision was that the building density should not be high and there should be sufficient land available to provide a rural belt between the town and other centres of population. Industrial development would be part of the town, so providing work and making unnecessary the travel commuting which was the inevitable result of the growing urban sprawl in existing cities. The Garden City Association was formed and two such towns were built in Hertfordshire: Welwyn and Letchworth (123). In each the factory area was apart from the town, but accessible. Building materials and styles were traditional. The houses had gardens and the civic centre was spaciously planned with tree-lined avenues.

Meanwhile, faced with an acute housing shortage after 1918 and rapidly deteriorating areas in inner cities of already bad housing, successive Acts of Parliament were passed to empower and assist municipal authorities to build housing estates at subsidised rents and to demolish the worst of the slums which had been created in the first place because of the inability of the tenants to pay an economic rent to the private enterprise operator (124). So existing towns spread quickly outwards and transport facilities were extended to ferry commuters from city centre to suburbia.

The destruction of town housing during the Second World War, together with a standstill in construction for its duration, had created another great shortage of accommodation by 1945. The 'new towns' and municipal city development of the succeeding 30 years were the equivalent of the garden cities and council housing of the earlier era. This time the use of newer materials

120 *Making sheet (cylinder glass) in the eighteenth century. Splitting open the cylinders*

121 *Making plate glass in France in the eighteenth century. The molten glass is poured on to the table and rolled out evenly by a copper roller.* Diderot's Encyclopaedia, *1763*

122   *Twenty-two storey block containing 640 flats, Lambeth, London, 1966. Concrete on-site precasting by Wates System*

123   *Terrace houses in Welwyn Garden City, by Louis de Soissons*

124   *Page Street Estate, Westminster, 1928. Grey brick and Portland Stone, by Sir Edwin Lutyens*

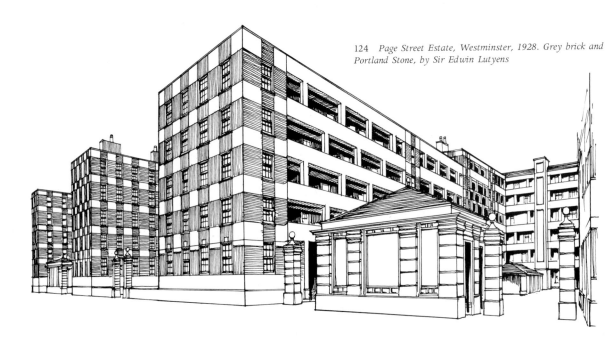

and methods of construction (pages 103, 106) cut the costs and increased the speed of erection. Gardens were smaller or gave way to open-plan design and architecture became standardized, but homes possessed a standard of comfort and convenience not hitherto dreamt of.

High-rise building

The skyscraper was conceived and named in America where, in the 1880s, conditions were ideal for its creation. Steeply-rising land values provided the incentive to build high; cheap production of iron and steel in quantity provided the material; Elisha Graves Otis (1811–61) provided, in his adaptation of the hoist for passenger use, the means of access to the upper floors. As long as people had to climb stairs buildings were limited to a height of five to six storeys. Various types of hoist to raise and lower loads had been known to man since ancient times. These had been powered successively by human hands, animals, water and, eventually, steam, but were employed only for freight because of the danger inherent in the possibility of the rope breaking and so causing the platform to fall. Otis' contribution was a safety device which he invented in 1852. In this, if the rope broke, springs forced pawls* on the lift cage to engage the ratcheted guide rails fitted into the sides of the shaft and so stopped the lift. In 1854 in New York Otis personally demonstrated his device when he cut the rope and stepped unscathed out of his lift.

With the development of the passenger lift taller buildings reaching 10–12 storeys, constructed by traditional methods using traditional materials, were erected in New York by the late 1870s. But it was difficult to build higher than this as the structure would require immensely thick walls on the lower storeys to carry the weight above. The logical answer to the problem was to construct a steel skeleton to which the walls, then no longer load-bearing but simply a cladding incorporating the windows, were hung. The process of development of such metal framing was gradual but, once established, quickly led to the multi-storey skyscraper. An early landmark in this development was William Le Baron Jenney's (1832–1907) Home Insurance

Building in Chicago, erected 1883–5. Using iron columns, lintels and girders with steel beams, he produced a load-bearing metal framework, structurally independent of the outside curtain walling. This was followed by the fully developed steel skeleton construction of the Tacoma Building in the same city (Holabird and Roche) and, by 1892, the 21-storey Masonic Building. Other American cities soon followed Chicago's example, notably New York where the Woolworth Building of 1913 reached 60 storeys.

In Britain, by 1930, congestion in large cities had become so acute that it was realized that accommodation in flats represented the only way to solve the housing problem, but these flats were usually in long blocks of only five to six storeys each. It was the late 1940s before shortage and high cost of land in cities and ever-lengthening housing lists impelled municipal councils to build high-rise flats. Many councillors were attracted by the ideal of 'vertical garden cities' which would so economize the use of land that space would be available for open-plan gardens surrounding the buildings. Such schemes were thought of as a new design for living founded on the theories of modern architects like Le Corbusier and Gropius.

In support of this technical advances in steel-framing, the substitution of welding for riveting, the extensive adoption of reinforced concrete (page 108), an increased employment of mechanical plant on the building sites, wide-scale prefabrication (page 104), all contributed to the development of new building techniques suited to high-rise construction. The 1950s and 1960s were the decades when town councils erected most of the tower block estates (122). A decline in such building began in the late 1960s, partly attributable to the dramatic collapse of one corner of the 22-story system-built block of Roman Point in the London borough of Newham (1968), but also because some of the social implications of this type of urban living were slowly beginning to be appreciated, particularly with regard to the elderly and people with young families.

Prefabrication

The idea of making up in a factory building units

---

* A bar or lever fitted with a catch to engage the teeth of a ratchet.

which comprise sections of walling, roofing, window and door openings and which could be assembled on site and affixed to a skeleton structure was experimented with in the nineteenth century: the Crystal Palace of 1851 (page 98) was the most famous example. Impetus was given to adopting systems of prefabrication by the shortage of housing accommodation occasioned by the First World War, and, during the years 1918–39, several ways of using these methods were tried out, especially between 1918 and 1925, when there was a shortage of bricks. Steel window frames were standardized to fit walling units, then wooden frames were similarly adapted. Large pre-cast concrete panels were made, suited to building walling and roofing at speed. Steel-framed houses were built, the structure being designed to take standardized panelling. After 1925, as the brick-making industry became re-established, brick was once more seen to be the preference of the British family and the prefabrication method languished.

An even more acute shortage of accommodation in 1945 brought back the prefabrication concept and the government sponsored a number of firms and schemes to erect temporary houses which became popularly known as 'prefabs'. Several principal designs were produced by groups of architects in combination with construction firms and suppliers. Arcon, which took its name from the firm of architects which designed the houses, was one of these, the project being undertaken by Taylor Woodrow Construction Ltd. The Arcon single-storey house consisted of some 2,500 parts made by 145 different manufacturers. It was designed to be delivered complete to the site from storage depots in four lorries, the first carrying the steel structure and panels, the second the cladding and wooden floors, the third the internal partitioning and the last the kitchen/bathroom unit, finishes and trim. The construction was of light steel-framing clad externally with asbestos-cement sheeting and lined internally with plasterboard; floors and internal doors and framing were of wood. Thermal insulation was good. Accommodation was not spacious, but the cooking, storage, heating and washing facilities were an immense advance on any previous municipal housing. Of particular

interest was the kitchen/bathroom unit, which was planned with the service fittings of the kitchen – sink, cooker, wash-boiler and refrigerator – placed back to back with the service fittings of the bathroom, so making it possible to contain all the plumbing in one transportable unit (125).

When the post-war housing emergency was over, again prefabrication was less widely employed, but in the 1970s, as building costs rose sharply, a more extensive use of factory-made units took place and standardization, the essential concomitant of the prefabricated method, became fully effective. The standardizing of separate building parts had been vital from the earliest experimentation, but a system to establish an overall three-dimensional-unit of measurement had been slow to be accepted in Britain. Such a system is known as modular design and ensures the accurate fitting of all building parts, whether bricks, panels, doors, windows or furniture, no matter where or by whom manufactured. The Modular Society* was formed in Britain in 1953 to promote such a system of uniformity, its membership comprising all concerned with the building industry from architects to clients and craftsmen.

Building materials

Notable advances were achieved from the late nineteenth century onwards in increasing production whilst reducing labour costs in the making of traditional materials for the construction industry. In response to accelerated demand the methods of manufacturing bricks were revolutionized. In this the development of the Fletton process in the 1890s was of particular importance. The name refers to the shale-clay found in ample quantity at Fletton on the outskirts of Peterborough, which possesses a low degree of pliability due to a reduced water content and so can be pressed into a brick which could be fired without being previously dried. Also, less coal is required to fire the clay as it already includes approximately ten per cent of carbonaceous material. The brickmaking industry steadily became more mechanized during the

* The work of the Society is now handled by the British Standards Institution which deals with the problems of the industry.

125 *Arcon Mark II prefabricated house first erected 1944. This front elevation shows the cantilever porch. The boxes are for fuel and refuse*

126 *Floor slab being lowered into position on a low-rise block at Highfield, Feltham, 1960s. Concrete on-site precasting by Wates System*

twentieth century. This applied to the digging out of the clays, separating the different strata, crushing and grinding the material, moulding, drying and firing. All processes were streamlined and accelerated.

In 1900 wood was still reasonably cheap, but its cost rose sharply as the century progressed so that, since 1945, solid timber has been used in building only where absolutely necessary. On the other hand, a wide range of wood products have been developed as more sophisticated machinery and better adhesives have become available. A patent for plywood was taken out as early as 1840 and during the nineteenth century plywood was used as a cheaper substitute for solid wood for furniture and interior fittings (page 130). The wood was cut by saw and the thin sheets glued together, each layer being placed so that the grain ran at right angles to that on the previous one. Laminated timber, used for structural purposes, was made so that the grain of each piece ran in the same direction and the layers had to be bolted or screwed together. The advent of efficient knife-cutting machines which took advantage of being able to cut round a log so as to provide a continuous layer, and also the introduction after 1945 of synthetic resins, have produced cheaper, stronger plywood, an important material in its own right. Particle board is another product, made from chips and shreds of wood compressed with synthetic resin.

Window glass manufacture has undergone tremendous changes since the First World War. Chance Brothers of Birmingham first developed a system of passing molten glass through rollers in 1887 and from this the continuous rolling process gradually evolved in which a perpetual band of molten glass is drawn, day and night, from immense tank furnaces heated by oil, gas or electricity to temperatures of 1,300–1,500°C. Sheet glass was first produced in this way and, after 1923, plate glass also, but the latter still had subsequently to be ground and polished.

Then in the 1950s came the great technical breakthrough of the float glass process which produced, for the first time, perfectly flat glass, free from distortion and of great brilliance and clarity, yet needing no grinding or polishing.

This remarkable process was developed commercially at Pilkington's in St Helens, Lancashire in 1959. In it the molten glass floats in a continuous ribbon along a bath of molten tin, leaving the furnace at a temperature of 1,500°C and held at 1,000°C by a chemically controlled atmosphere until the surfaces have become flat and parallel so that the ribbon of glass has cooled sufficiently to be removed without the surface being marked. Over this temperature range the tin remains molten yet is sufficiently dense to support the glass (127).

Among the selection of special glasses now available for domestic use is solar control glass, which is intended to reduce excessive heat from the sun and protect furnishings from fading. This is slightly tinted, either with a sun-reflecting coating added to the glass surface in manufacture or the product is laminated and contains a tinted interlayer. Replacement double-glazed windows, factory-made to fit existing openings, are being increasingly installed in homes to reduce draughts, noise and fuel bills. In these sealed units the space between the two layers of glass is flooded with nitrogen.

Chief among the new materials characteristic of twentieth-century domestic building are concrete, asbestos sheeting, aluminium, plastics and fibreglass. Concrete is the oldest form of a man-made building material; it was utilized as a mortar in ancient Egypt and in ancient Greece, then developed by the Romans as a structural material. Concrete* is composed of four major ingredients: sand, stone, cement†, water. The problem in making an enduring substance lies in the type of cement employed. The Romans added *pozzolana* (page 98) (which resembled a reddish sand) with lime to make their concrete. After the collapse of the western half of the Roman Empire in the fifth century, much of the knowledge and experience which had been gained in making and utilizing concrete was lost for about 1,300 years; it was used only as a mortar.

It was not until the middle of the eighteenth century that builders began to search for the

---

*The word derives from the Latin *concretus* = grown or run together.

† From the Latin *caementum* = rough stones, rubble, building material.

127 *Float glass plant automatic warehouse, Pilkington Brothers Ltd, 1963. Float enters from right of picture*

128 *Four-inch thick fibreglass Supawrap 100 being laid between the joists of a domestic attic*

means to make a cement which would set quickly and be strong and durable. *Pozzolana* and lime mixtures had been imported from Italy since the sixteenth century, but the secret of Roman concrete had not been understood. John Smeaton achieved the first major success (page 98). Further research and experiment by others, notably the French engineer Louis J. Vicat (1786–1861) showed that Welsh limestone such as that used by Smeaton contained (as *pozzolana* had done) silica and alumina.

During the nineteenth century many advances were made in providing better cements, for example, Aspdin's Portland Cement, which he so-named in 1824 because he thought the colour resembled that of the famous limestone, and Johnson's better version of 1845 (page 98). Improvements were also made in kiln design to produce a cheaper cement more quickly and efficiently.

The developments which finally made concrete into the ubiquitous material of the twentieth century were its reinforcement with steel and its pre-stressing. Concrete used by itself is very strong in compression, that is, it resists the weight of a load which tries to squash it. But it has poor tensile strength, that is it cannot resist forces which threaten to pull it apart. It is, therefore, unsuited for use in arches, domed roofs or structural beams. The Romans who employed it for these purposes built so massively (walls up to eight feet in thickness) that their vaults could take the strain. If, however, concrete is reinforced with metal rods or wires, the elastic strength of the steel will absorb all tensile stresses and complement the high compression resistance of the concrete to provide an immensely strong and cheap constructional material. Experiments were made during the nineteenth century and by the 1890s bridges were being built of steel-reinforced concrete. Pre-stressed concrete was pioneered by the French engineer Eugène Freyssinet (1879–1962) in the years after the First World War. The theory behind this process is that if the metal reinforcement is stretched before the concrete is poured into position and the pull maintained until the concrete is hard, it will prove a much stronger and more durable material.

Concrete was not widely employed for walls and roofs in domestic building before about 1930, when a number of firms began to manufacture pre-cast panels and blocks for this purpose. Then, again in 1945, the suitability of this material for casting into all kinds of shapes needed for the immense crash building programme in prefabrication was fully realized and concrete was thereafter extensively and increasingly used for all types of housing from steel-framed blocks of flats to low terrace structures. In the 1950s and 1960s some constructional firms, notably Wates Ltd., developed a system of on-site precasting of concrete which economized in transport and handling as well as skilled manpower. A flexible system of large precast panelling with interlocking wall and floor elements was evolved in the Wates System, especially suited to the erection of high- and low-rise blocks of flats (122, 126). Breeze blocks, made from concrete containing an aggregate of coke furnace ashes, were and have continued to be employed as a cheap and easily-made material for partition walls and the inner skin of cavity walling.

Asbestos is an incombustible fibrous material which was spun and woven into fire-resistant fabrics as early as 1870. The fibres which were too short for spinning were then thrown away, but by 1900 it was discovered how to make an asbestos cement with these by mixing them with hydrated inorganic cement in a proportion of 15 per cent asbestos fibre and 85 per cent cement. This new building material then came to be widely used to make corrugated sheeting, roofing tiles, wall boarding and, later, guttering and piping. In recent years, as the health risks to workers in the industry have become more apparent, essential safety precautions have made manufacturer more costly so leading to a decline in the use of asbestos and its consequent replacement by plastics and fibreglass (pages 83, 81).

Aluminium had been utilized in the home for cookware since the early years of the twentieth century (page 70), but its application as a building material came more slowly. Aluminium window frames began to replace steel designs (which required painting) in the 1930s, but these were expensive. Only after 1945, with the development of better extrusion processes, did the metal come into wider use both structurally and as cladding and framing.

The potential application of plastics to structural needs has not yet been fully developed though experimental work and manufacture points to their satisfactory employment in combination with other materials. A notable instance of this is the reinforcement of certain plastics with fibreglass (page 83). Aside from load-bearing structural units, different plastics are being utilized in ever-increasing quantity in every aspect of housing as in, for instance, cisterns, piping and roofing panels (page 81).

GARDENING

In the early nineteenth century the middle classes began laying out their gardens in suburbia and by 1830 gardening magazines started to appear. Soon garden equipment was offered for sale. Today all kinds of machines and gadgets are available for easy-care gardening but now, as in the 1830s, grass cutting is probably the principal chore. The first lawnmower was patented in 1830 by its inventor, the engineer Edwin Budding. In its design he had adapted a machine for cutting pile on cloth to a larger version for grass. A similar machine was exhibited at the 1851 Exhibition (129). By 1910 the lawnmower was still pushed by hand (130), but before too long petrol-driven mowers were available and later still designs driven by electric power. In 1963 the Flymo (short for Flying Mower) was designed by the Swedish engineer Karl Dahlman. A light, easy-to-use cutter which can be moved in all directions, the Flymo floats on a cushion of air on the hovercraft principle (131).

130 *Excelsior lawn mower. Catalogue 1910.
Price approximately £4.50 (14-inch)*

131 *19-inch GL Petrol-driven
Flymo lawnmower, 1979*

CHAPTER SIX

The Home 3
The Interior

SURFACE COVERINGS AND DECORATION

Walls

Since the early Middle Ages several methods have been in use in Britain. At first the most common, particularly in well-to-do homes, was to paint the plastered surface in pleasant colours with decorative patterns or varied scenes. Alternatively, hangings of wool, silk, brocade or velvet gave warmth and decoration and, in wealthy households, tapestries were hung.

In the fifteenth century joined framed wood panelling was introduced from Flanders to line walls and give warmth and richness to interiors: this was an important technical advance. The oak panels were tapered on four sides to fit into grooves made in a framework of horizontal and vertical members which were mortised and tenoned together (page 128). This panel and frame construction (made by a craftsman joiner) was more satisfactory than earlier work made by a carpenter; it allowed space for expansion and contraction and so maintained its form. The panels of early work, which only covered the lower part of the wall, were made of overlapping vertical boarding (as in a clinker-built boat), but by the end of the century linenfold panelling, carved to resemble folded material, covered the whole wall up to frieze level (132). Elizabethan panelling was more ornamental, carved or inlaid. Panelling was widely employed during the seventeenth century when, in current architectural fashion, the room was often designed in the form

of a classical order, doors, windows and chimneypiece being incorporated into the scheme. Curved bolection mouldings provided a convenient and attractive means to raise the panel surface above that of the surrounding framework.

Wall hangings made of paper (wallpaper) were first made to be less costly substitutes for the rich patterned tapestries and embossed leather hangings adorning the wealthy homes of the sixteenth and seventeenth centuries. With the development of letter-press printing (page 154), wood blocks were also made to print line designs on small squares of paper, colour then being added by hand. No attempt was made to make the patterns of the different squares match or create a complete design. Many such papers were then applied as decoration to wooden panelling. The oldest example of such block-printed wallpaper was found in 1911 at Christ's College, Cambridge. The work of Henry Goes of York, it is a black and white design printed on the back of a bill proclaiming the accession of Henry VIII (1509) (135).

Gradually the making of wallpaper panels was developed and improved. Colours were printed, using successive blocks and designs matched to give an overall pattern. In the 1760s in France wallpaper was first made in long strips by pasting the sheets (each 12 by 16 inches) together before printing, giving a roll 12 inches wide and 32 feet long. In France in the 1620s a method of making flock papers was invented which closely imitated

132   Oak panelling, early sixteenth
century. Carved in linenfold and
Renaissance ornament

133   Wreath wallpaper.
William Morris, 1876

the rich and costly Italian cut velvet hangings. In this a sticky varnish or glue was printed on instead of coloured paints. While it was still wet, finely chopped coloured wool was blown on to the surface using a small bellows and the fragments stuck to the design. Later, silk was used which made the resemblance to cut velvet more realistic. Flock papers became popular in England where they were manufactured by the 1630s though if the quality of the paper available was not strong enough, linen or cotton was used.

Hand-painted Chinese papers (also in small sheets) were available in England from the early seventeenth century (134). These were expensive but in great demand in order to blend with the then fashionable Oriental porcelain and lacquered furniture (pages 86, 128). Gradually English printed imitations of such papers began to be made and, later in the eighteenth century, John B. Jackson manufactured in Battersea a variety of printed papers in floral patterns and versions of paintings by masters such as Canaletto and Poussin.

By the 1830s patents were being taken out for wallpaper printing machines, but it was a further decade before successful results were obtained, after which machine-printed papers began to replace hand-printed ones. Paper in continuous rolls was produced in France in 1799, but its use was not permitted in England until 1830 because of the high revenue derived from the tax levied on the small sheets. Early wallpaper panels had been nailed or pasted to the wall, but the costly flock and Chinese papers were carefully protected from damp and dirt by being mounted on wood frames backed by linen and paper. With machine printing, papers in rolls became less expensive and could once more be pasted to the wall (133).

Ceilings and walls

During the Middle Ages wood roofing was most usual, either in open timber structures or flat ceilings with cross beams and boarding between. By the sixteenth century plaster infilling was replacing the boarding and in Elizabethan times ceilings and friezes were decoratively plastered.

The ordinary plaster in use for interior work was generally made from lime, sand and water with various substances (animal hair, dung, blood) added to help to bind the substance and avoid cracking. Gypsum plaster (called plaster of Paris because of its French origin) was available in the Middle Ages, but its use was reserved for fine finishes in well-to-do homes because of its higher cost. The ornate decorative ceilings of the sixteenth, seventeenth and eighteenth centuries derived from the Renaissance plasterwork initiated in fifteenth-century Italy when Italian craftsmen experimented with a type of plaster which had been used by the Romans, one which was malleable and fine yet set very hard: they called it *stucco duro*. It contained lime and some gypsum but also powdered marble. Craftsmen in Elizabethan England adopted both stucco and the Renaissance forms of decoration and incorporated empirically a variety of other additions, for example, milk, eggs, ale and beeswax. From the later seventeenth century stucco decoration was used more and more on walls also and gilding and painted panels were included in the decorative schemes of all surfaces.

With nineteenth-century industrialization it became more usual for the architect or builder to purchase decorative motifs and mouldings by the yard and fasten these to ceilings and friezes rather than work *in situ* as before. Plaster was applied to an underceiling of laths and plaster to give a plain surface to which the ornament could be affixed. As a result of the First World War not enough plasterers were available and plaster board began to be employed as a replacement material. These plaster panels then consisted of a layer of gypsum plaster between sheets of strong paper. Though used extensively in the USA by 1910, in England trade union opposition delayed its general adoption by builders until nearly 1930. It was the need for house building and repair after the Second World War, as well as an even more acute shortage of plasterers, which led to the widescale use of plasterboard. By this time the product had been greatly improved to conceal joins, to be suitable for direct application to brick and concrete surfaces and to provide better thermal insulation by adding a backing of aluminium foil.

Floors and other horizontal surfaces

Floor surfaces of stone or brick blocks, tiles or polished wood boards were partially covered by

134   *Panel of Chinese wallpaper. Part of a continuous picture for a room. Hunting, boating and festival scenes. Eighteenth century*

135   *(Top left) Facsimile of a portion of wallpaper found at Christ's College, Cambridge. Wallpaper date c.1509*

136   *Formica decorative laminates employed on kitchen surfaces in 1981. Woven grass pattern on cupboard doors and tile pattern on worktops and walls*

plaited rush mats from the sixteenth century; imported carpets were still rare and were deemed too valuable to be placed on the floor. Inlaid patterned wood floors were fashionable in well-to-do homes by the seventeenth century and small rugs were laid on them. It was after 1750 before English carpet manufacture began to make it economically possible to cover a large area of floor with carpeting (see carpets, page 149).

Easy-care floor coverings to be washed or polished came much later. Linoleum was the first satisfactory answer and represented a great breakthrough in alleviating the cycle of scrubbing of tiles, bricks and wood. It was the Englishman Frederick Walton who first set up a factory at Staines in 1864 to make a linoleum (not dissimilar to a twentieth-century one) which consisted of a burlap base coated with a cement made from linseed oil, gum, resin and colour pigments. Linoleum was made in rolls in ever-increasing quantity until about 1950, by which time it was being replaced by synthetic coverings in roll or tile form of materials such as vinyl (page 85).

The working surfaces of the kitchen which, for centuries, had been scrubbed wood, were revolutionized by the development of decorative plastic laminates which could be applied as a veneer one-sixteenth of an inch thick to a wide range of core materials – wood, particle board, plasterboard, etc. – to provide an easy care, hard, resistant surface for all kitchen (as well as bathroom, toilet and nursery) needs. The household name in this field is Formica laminate, now produced in an extensive range of designs and colours. It was in 1913 that a method was found in the USA to make a laminated plastic sheet and decorative laminates were being manufactured in the 1930s, but the main acceleration has been since 1945. In the process of manufacture numbers of layers of paper, impregnated with synthetic resin, are subjected to considerable pressure and heat (136).

LIGHTING

Oil and candles

A level of illumination which enables everyone to continue their domestic occupations as well during the hours of darkness as during those of daylight is one which (though now taken for granted) is recent indeed. In Britain for two millennia the only artificial light available came from candles and oil lamps and, for much of the time, the level of illumination was very low, constant attention needed to be given to trimming wicks and the burning fat dripped and produced a smoky, smelly atmosphere.

The cheapest form of domestic illumination was the home-made rushlight, made by repeatedly dipping a dried, peeled rush into tallow (derived from animal fat), which had been melted in a greasepan. Candles were also generally likewise home-made from tallow poured into moulds into which the wicks had first been inserted. Beeswax candles burned brighter, needed less attention and smoked less but were costly. With the expansion of the fishing industry in the late eighteenth century sperm whale oil began to be used to make candles, and when, in the following century, paraffin wax was produced from petroleum, a blend of the two materials produced a greatly improved flame. Still more advantageous was the later stearine candle which was firmer and gave a brighter light without the accompanying acrid odour. This development owed much to the work of the French chemist Michel Eugène Chevreul whose researches between 1811 and 1823 demonstrated the chemical nature of fats (138, 139).

Simple oil lamps made of earthenware or metal were in use from Roman times in Britain. These were containers for the fuel and a floating wick with adjustment for the frequent trimming needed to prevent smoking (137). Animal and vegetable oils were burned, particularly olive oil, rape seed (colza) oil derived from kale and, from the late eighteenth century, whale oil. After paraffin became available in the 1860s this gradually replaced other fuels.

It was Ami Argand, the Swiss chemist, who in 1784 made the first significant contribution towards improving the illumination of the oil lamp which, up to that time, gave only one candlepower for each wick. Argand devised a tubular wick in the form of a hollow cylinder which he enclosed in a chimney (first metal, later glass), thereby inducing a current of air inside

137　*Roman earthenware oil lamp*

138　*Candle snuffers, 1547–53*

139　*Silver-gilt chamber candlestick and snuffer*

140　*Colza oil burning Argand reading lamp. The burner is fed by oil flowing under gravity from the reservoir above the level of the wick. The oil level in the reservoir is controlled by a float valve which closes the filling hole*

141　*Paraffin lamp with duplex wick, c.1890*

142　*Carcel forced-feed oil lamp*

and out which so improved combustion that the light from a single wick was increased up to ten times (140). The next advance was to increase the flow of oil which, in the early Argand lamps, was by means of capillary action or by gravity; the latter method meant placing the oil reservoir above the wick, but this cast a heavy shadow. The first successful mechanical lamp with spring-driven pump to supply fuel under pressure from a base reservoir to the wick was produced in 1800 by the Frenchman Carcel, and another Frenchman Franchot patented a similar but simpler and cheaper lamp in 1836, known as a Moderator which became very popular (142).

Twentieth-century paraffin oil lamps (141) give a brighter light because the fuel is vapourized before it reaches the flame and so less energy is wasted; this is achieved by heat and pressure. In the modern Tilley pressure vapour lamp, for example, pads soaked in methylated spirits are clipped to the centre tube and set alight. The control valve is opened and the fuel pumped up from the reservoir. The paraffin in the tube is vapourized by heat from the pads and the mantle ignites.

Gas

Twenty-six miles of gas mains had been laid in London by 1816 and soon gas illumination of the streets and public buildings was acclaimed a great success (page 32), but domestic consumers were less enthusiastic. Even in 1850 the standard burner (batswing or fishtail design) flickered and smoked; it contained sulphur compounds which gave out an unpleasant smell and the fumes damaged the furnishings as well as dirtying the interior decoration (144).

The discovery which revolutionized domestic gas illumination, giving an infinitely better light while, at the same time, doing away with much of the dirt and smell, was that of the gas mantle. Attempts to make such a mantle had begun by 1840, but until the invention of the bunsen burner in 1855 (generally attributed to the German chemist Robert Wilhelm von Bunsen 1811–99), it had not proved possible to produce a gas flame of sufficiently high temperature. In the bunsen burner air was mixed with the gas before ignition so giving a hotter flame.

The incandescent mantle was devised by the Austrian Carl Auer von Welsbach (1858–1929) in 1886 (146). In it he was making use of the principle that when the temperature of a substance is raised sufficiently it begins to glow and so emits more of its energy in the form of light. Drawing upon his knowledge gained by his research into rare earths, he made a mantle of knitted cotton impregnated with solutions of such rare earth oxides. When the cotton mantle had burned away the skeleton of the material retained its form. He found eventually that the best mixture for his mantle was 99 per cent thorium oxide and one per cent cerium oxide. The Welsbach mantle went into production in 1887 and improved, less fragile versions followed in 1890. Another important advance was that of the inverted mantle in 1903, which directed the light downwards and eliminated shadows caused by the fitting (143).

Electricity

The arc lamp was the earliest form of electrical illumination. Its use was based on power from the voltaic pile (page 35) and as early as 1802 Sir Humphry Davy drew attention to the vivid light emitted by the electric spark struck between two carbon electrodes (page 70). More than half a century elapsed before much use could be made of this knowledge which had to await the availability of an adequate source of electrical power which came with Gramme's ring dynamo of 1871 (page 35). Soon arc lighting was being installed for use in factories, streets and railway stations. In England R.E.B. Crompton (1845–1940) designed one of the early arc lamps and in his factory at Chelmsford manufactured them. In 1879 he was responsible for illuminating the Henley Regatta and later the grounds of Alexandra Palace with Crompton lamps powered by Gramme generators. But the arc lamp was unsuited for domestic illumination; the light was dazzling, too costly and needed constant adjustment – pieces of very hot carbon were liable to break off. Electric light in the home had to await the filament lamp.

The concept of the incandescent filament lamp had for years been believed by several researchers in the field of electrical illumination to

143   *Metro inverted burner with incandescent gas mantle, c.1910*

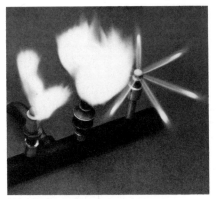

144   *Early gas burners*

146   *Welsbach type C gas burner with mantle, 1893*

147   *Edison-Swan carbon filament electric lamp in a wooden holder fitted to a gas wall bracket (globe omitted)*

145   *Early experimental Edison electric lamp, 1880*

149   *Sir Joseph Wilson Swan*

148   *Swan experimental carbon pencil lamp, 1878–9*

be the answer to domestic needs. The principle was the same as for the incandescent gas mantle (page 117), but there were three main problems in manufacture: to make a filament which was durable, to evacuate enough air from the bulb which contained it and to provide a satisfactory seal to the wires carrying the electric current as they passed into the bulb. Solving these problems took over 30 years from the time when W.E. Staite first demonstrated his incandescent filament lamp at a lecture in Sunderland in 1847. His lamp was short-lived as there was too much residual air in the vessel and the filament, made of an alloy of platinum and iridium, disintegrated. In Staite's audience at the lecture was the young Joseph Wilson Swan (1828–1914) who, following up Staite's work and J.W. Starr's use of carbon filaments in 1845, made his own first carbon filament lamp in 1848. This too was short-lived and it was not until after the invention of the mercury vacuum pump by Hermann Sprengel in 1865 that further research became feasible.

Eventually Swan (149) made his first effective carbon filament lamp in 1878 (148). By 1880 he had produced a more successful filament by treating the cotton thread with sulphuric acid before carbonizing it. Meanwhile the American inventor Thomas Alva Edison (1847–1931) had made his incandescent lamp using a platinum alloy filament. Edison continued to experiment with filaments of other materials and produced a carbon one in 1879 using carbonized paper and, later, one of bamboo (145). By 1882 carbon filament incandescent lamps were being made in quantity, Edison lamps in America, Swan lamps in England. Despite their high cost they were an immediate success because they were clean, safe and convenient. Only the gradual wiring of all homes for electricity, which took many years, held back domestic electrical illumination.

The carbon filament lamp dominated the market until about 1910. There were several companies manufacturing lamps in the early years and there arose intense rivalry between them. Edison was a better business man than Swan and his products became more widely known in the USA and Europe (only in England was Swan acknowledged as the inventor) as Edison had been quicker to patent his contri-

butions and more efficient at marketing his lamps. In 1883, though on the verge of litigation, both companies pulled back from the brink, realised that their advantage was to co-operate and they merged to become Edison and Swan United Electric Company (147).

The carbon filament lamp was not very efficient and in the first decade of the twentieth century extensive experimentation took place to find a metal filament which would withstand the high temperatures needed. In 1898 von Welsbach (page 117) produced the first satisfactory metallic filament of osmium which went on sale in 1902. Tantalum filaments followed and then in 1906 tungsten was first used. An improved, ductile form of tungsten filament was marketed in 1911 and this material, with its high melting point of 3,410°C, is still in use. One problem with such high temperatures is that the metal evaporates and blackens the bulb. Research at the General Electric Research Laboratory in the USA under the American chemist Irving Langmuir (1881–1957) led, in 1913, to solving this difficulty with the development of the much more efficient argon-filled incandescent lamp with coiled filament. Later the coiled-coil filament was developed*. Today's modern lamp gives about four times as much light as a carbon filament one for the same consumption of electricity. Other improvements have included the introduction of interior frosting of the bulb to reduce glare (1925) and, more recently, the addition of a fuse to avoid explosion of the bulb which might take place when a filament burns out.

A more efficient form of electric lighting developed in the 1930s is the discharge lamp which gives much more light than a filament lamp for the same consumption of electricity. As long ago as 1858 the German inventor Heinrich Geissler made a thin glass tube in which he obtained a glowing light by discharging electricity through mercury vapour. In the early twentieth century further experiments improved the Geissler tube using a variety of different gases, and by the 1920s neon, in particular, was

Coiled filament   Coiled-coil filament

being employed for advertising signs. In the 1930s the two best-known versions of these discharge lamps, as they were called, came into general use for street lighting: these were the mercury vapour lamp (bluish light) and sodium vapour lamp (yellow).

Discharge lamps are not of practical use domestically, but the fluorescent lamp, which developed from the mercury vapour discharge lamp, is now increasingly fitted in the home, especially in bathrooms and kitchens. Fluorescent lamps, introduced in the late 1930s, are coated on the inside by several different phosphorescent compounds (phosphors). The ultra violet light which is produced by the discharge of electricity through the mercury vapour in the tube is absorbed by the phosphors which emit light. The colour of such lamps can be altered by using different types of phosphors. Fluorescent lamps use little electrical power and the tubes last much longer than the equivalent filament bulb.

HEATING

Solid fuel

Until the early seventeenth century, when diminishing timber supplies gradually forced people to burn coal, logs blazing on an open hearth were the usual way to heat a room (see cooking, page 60). Alternative fuels were peat, turves and heather. Supplementary heating was provided by charcoal burned in a portable, metal brazier. This could be dangerous in small, ill-ventilated apartments: it was not unknown for sleepers to die quietly while breathing carbon monoxide fumes given off by the burning charcoal.

With the adoption of coal as primary heating fuel the chimney and hearth design changed and metal grates were devised to hold the coal. Living room grates followed a similar pattern to those described for cooking (page 60), but were more elegant and decorative and made from finer materials (150). The Victorian fireplace was generally of cast iron with steel or brass decoration. The chimney draught was more carefully controlled by reduction of the firebasket area and provision of smoke canopy and chimney register (a metal sheet which could be slid across the flue

opening to adjust the air current). The modern open fire, where still in use, burns smokeless fuel in an efficient, economical manner. Convection heating and provision of hot water may be incorporated. Gas or electric lighting devices are usually fitted. An alternative is the free-standing or fitted stove which controls combustion better and so wastes less heat. Victorian stoves were generally ornate, of cast iron (151, 154); modern ones are often finished with coloured, glazed enamel (153).

Paraffin heating stoves

When paraffin became available in the 1860s (page 33), these heaters were made in large numbers, chiefly for use in rural areas where homes had no gas (or, later, electricity). Victorian stoves were ornamental, in cast iron, and early versions were primitive, inefficient and odorous. The Valor Oil Stove of the 1920s was a great improvement, the paraffin being contained in a removable brass vessel on which stood the wick holder with circular wick (152). The traditional design has changed comparatively little since.

Gas

The same drawbacks attended early attempts to utilize gas for space heating as applied to cooking: it was expensive, it was smelly since the products of combustion were not properly disposed of, and it was inefficient (pages 32, 62, 117). The best of the early designs of the 1830s were convector types of stove which heated air by applying a flame to the exterior of a tube through which air passed.

After the invention of the bunsen burner (page 117), many attempts were made to design gas fires on the radiant principle in which the gas flame would heat a material to incandescence, but the problem was to find a suitable material. A variety was tried – pumice balls, fire-brick, woven wire – but the first real success was obtained with tufts of asbestos stuck into clay fire-brick; the gas flames heated the clay and the asbestos tufts became red hot, giving out heat with a cosy brightness (159).

After 1882 gas fires began to be sold in large numbers. The columnar radiants were introduced in 1905 and the more familiar grid pattern of columnar fire-clay radiant in 1925. The quality

50 *Cast iron hob grate from 15 Portman Square, London, c.1790*

151 *Iron Tortoise stove. Catalogue, 1910*

152 *Valor paraffin stove, 1928. Shown open in preparation for filling and lighting*

153 *Courtier room-heating stove by Esse, c.1955*

54 *Cast-iron anthracite stove. Catalogue, 1910*

of radiants was steadily improved and automatic methods of lighting were developed.

In the early twentieth century a new generation of convector heaters was put on the market (155). Characteristic was the Embassy twin-column radiator of 1920 which contained two Argand-type burners. The effective convector gas fire of the modern type became available in the 1950s. In this air from the room was circulated round the burner and after heating was recirculated into the room.

Electricity
Heating by electricity, as with cooking, lagged behind its use for lighting (pages 32, 62, 117). The earliest heaters, put on sale in 1894 by Crompton and Co. Ltd (pages 39, 117), were electric radiant panels in which the heating wires were embedded in enamel in a cast-iron studded plate (157). They were not very successful because they broke easily due to the differing expansion (when heated) of the iron and the enamel. This type of heater was followed by a different idea put forward by H. Dowsing (a collaborator of Crompton) to use incandescent electric lamps for heating. The Apollo Luminous Electric Heater of 1904 was based on this idea. The cast-iron fire had four large sausage-shaped carbon filament bulbs backed by a reflector, but gave out little heat (156).

The two chief defects of these early fires were that the elements burned out frequently and a suitable means of wiring them had not been found. By 1914 these problems had been solved. In 1906 A.L. Marsh patented a resistance wire made from a nickel-chrome alloy which retained its strength even at high temperatures, and six years later C.R. Belling devised an element in which the wire was wound round a fire-clay former (158). These two developments revolutionized electric fire design, but, due to the high cost of electricty and the long time taken to supply power to households, it was after 1930 before electric heating became a serious competitor to solid fuel and gas.

The idea of electric storage heaters which would consume electricity at off-peak hours and retain the heat in concrete blocks to release it gradually in the intervening hours was initiated in the 1930s. It had only a marginal success, but was re-introduced in the 1960s when more efficient versions were made which contained bricks, also a fan to maintain a flow of air through the heater. Modern slim-line heaters have been further improved.

Central heating
Houses in Roman Britain were heated by hypocaust, a method in which hot air from a basement furnace passed through wall flues to heat all the rooms. The hypocaust was the underfloor chamber of brick or stone; the term derives from two Greek words meaning 'the place heated from below'.

After the fall of Rome the hypocaust fell into disuse and the idea of heating a building from one central source was forgotten until the introduction of steam power (page 27). Several methods were developed in the eighteenth and nineteenth centuries to use steam for this purpose, but though successful for factories and large buildings, steam was never seriously considered in Britain for home heating. Indeed, the wide-scale adoption of various means of domestic central heating has been a feature of post-1945 building. There is now a range of possibilities to choose from, powered from solid fuel, oil, gas or electricity. There are hot water systems with radiators, skirting heating or pipes embedded in the walls and warm air circulation convected through underfloor or wall grilles.

WASHING AND BATHING FACILITIES
The problems of heating sufficient water and its disposal, the unreliability of the domestic water supply and the inadequacy of sewage facilities all contributed to a marked reluctance to wash or bathe more often than absolutely necessary in the centuries before the Industrial Revolution brought solutions to these difficulties (pages 14, 96). An exception to this picture in Britain was when under Roman rule public baths with adequate water supplies, means of heating and drainage were available in most communities. Larger houses possessed their own bath-houses especially in rural areas.

When the Romans left, only the medieval monasteries set a standard in maintaining

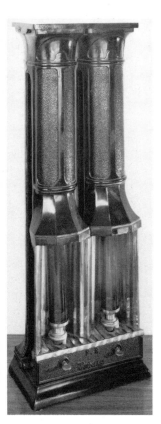

155 *Twin-column Embassy gas convection radiator*

157 *Crompton-Dowsing electric fires, 1893 from Crompton Catalogue*

156 *Apollo luminous electric heater, c.1904*

159 *Gas fire with ball fuel, c.1900*

158 *Magnet portable electric fire, 1925. Belling-type wired formers*

tradition of bodily cleanliness. Cold water was always available in basins or troughs for daily washing and water was warmed for regular (though infrequent) bathing; it was carried in ewers to wooden bath tubs.

There was little improvement in domestic washing facilities until the later eighteenth century. For most of the time water had to be carried from an outside pump and heated in a cauldron over the kitchen fire. People washed in the kitchen or their bedrooms. By the 1760s furniture designers were making purpose-built pieces for bedroom toilet use: toilet tables, shaving tables, washbasin stands, for example, all fitted with mirrors and toilet articles (160, 163). Small metal baths of different shapes became more common. The water supply was still intermittent, but with the introduction of the ball valve about 1750 the main house tap did not need to be turned on and off so often.

The great advances came from about 1850 onwards. Running water was supplied, at first only on the ground floor but later to bedrooms also, and then washbasins acquired taps and marble tops and were affixed to the wall (161). With better water supplies bathing became more generally accepted and various designs of bath appeared: the full or lounge bath (similar to a modern one) (169), the sitz (sitting) bath in which one washed the bottom half of the body only and, in a chilly room, could retain one's clothes on the top half, the hip bath (which was fitted with a seat so that only one's legs got wet) (164) and the slipper (boot) bath. This last design, made from sheet metal, was shaped like a shoe or boot. It was a sensible shape as it kept the water hot and only the bather's head and shoulders were exposed. More hot water could be added from above when desired (165).

The heated bath also appeared on the market. Charcoal or gas was used. The latter model wisely included advice to turn off the gas before entering the bath. In the 1880s, as more baths were equipped with taps and running water, bathrooms were fitted into many houses. These were large and ornate, with patterned tiling and elaborate fittings which included a shower, a bidet, towel rail and a washbasin (166). The bath was usually of cast iron, painted or galvanized.

Early in the twentieth century the more expensive porcelain enamel finish was developed. This was a vitreous coating which kept its quality surface well. It was finally replaced by the lighter-weight plastic modern bath.

Hot water for baths remained a luxury until the late nineteenth century. The boiling of water over the kitchen fire had been replaced by the coal- or wood-burning iron kitchener (page 60) or copper, but in both cases (before 1860) the water still had to be carried upstairs to the bath. So, apart from the heated bath, the gas geyser, invented in 1868 by Benjamin Waddy Maughan, was a godsend. Maughan christened his device a geyser, from the Icelandic word *geysir*, the name of a specific hot spring there, and meaning 'gusher' (171).

Other gas geysers followed, but for a long time it was hazardous to use them. Few of the early designs were fitted with flues, so the fumes were dangerous in small bathrooms. There was also no pilot jet on such appliances and the water supply was often unreliable. Until well into the twentieth century, if the user did not faithfully carry out the instructions for lighting in the advised order, an explosion could result. Eventually safety devices were incorporated which made gas geysers virtually foolproof. For instance, gas could not feed the main burner until the pilot jet had provided sufficient heat to operate the tap and, later, gas flow was designed to depend on water flow. Ewart became the household name in the manufacture of such geysers (170). Instantaneous gas water-heaters were also installed in the kitchen. The most familiar name here is Ascot. Heaters, later made in many sizes, were introduced from Germany in 1932 by Bernard Friedman.

Electric instantaneous water heaters also appeared on the market in the 1920s and electric immersion heaters were installed in lagged water tanks by 1930, but, both then and now, the chief drawback to this highly efficient and convenient method has been the high cost of electricity.

SANITATION AND TOILET FACILITIES

The room and the equipment which we now refer to as the 'lavatory' or the 'toilet' have been known by many different names over the years,

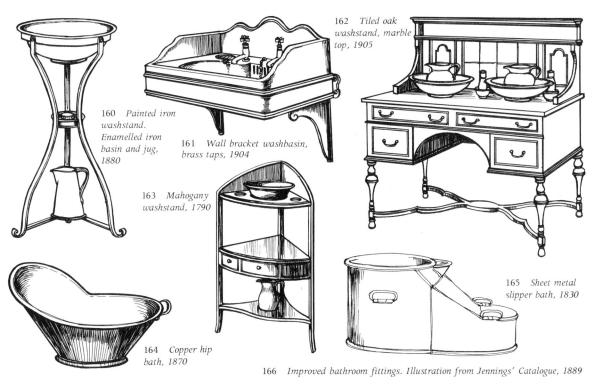

160 Painted iron washstand. Enamelled iron basin and jug, 1880

161 Wall bracket washbasin, brass taps, 1904

162 Tiled oak washstand, marble top, 1905

163 Mahogany washstand, 1790

164 Copper hip bath, 1870

165 Sheet metal slipper bath, 1830

166 Improved bathroom fittings. Illustration from Jennings' Catalogue, 1889

Spray and Plunge Bath.   Lavatory.   Dressing Table. Bidet.   W.C.

but, until the later nineteenth century, with very few exceptions, were all at best uncomfortable, unhygienic and malodorous and could be a dangerous hazard to health and even life.

To use the latrine or privy was to the Romans a communal, even social, activity and communities were provided with sanitary latrines, well equipped with running water and washing facilities, as can be seen in the extensive remains at Ostia, the port of Rome, where 20 marble seats, lining the walls, survive.

During the Middle Ages monastic houses had their necessarium built near or over a water supply; again, these were communal. Public latrines were provided in towns, generally constructed over rivers; London Bridge had a large latrine which served over 100 houses. In castles and manor houses the garderobe was built into the thickness of the great walls and drained into the moat.

Between about 1500 and 1750 sanitary arrangements deteriorated and the close stool and chamber pot largely replaced the garderobe. The close stool, used in well-to-do homes where it would be emptied and cleaned by servants, was a bowl within a padded, lidded box. Elaborate examples had arms so that they were like chairs to sit upon (168). Pots were emptied into the street. A large house might be equipped with a 'house-of-easement' which would be sited in a courtyard or basement. Such a structure would accommodate two or more people.

In the eighteenth century sewage disposal was still primitive and insanitary. The general method was a privy or closet outside the back of the house; in poorer quarters of towns one privy served many houses. These privies had wooden seats and were built over small pits. They had to be emptied by buckets, which were then carried through the house to larger communal cesspools, emptied regularly by night-soil men. There were many instances where the pits drained through into the drinking water supply, with serious results. Indoors, chamber pots were used. They were often concealed in pieces of furniture, the sideboard for dining room relief of gentlemen after the withdrawal of ladies and, in the bedroom, in a night table or night commode (172).

The invention which eventually abolished the malodorous germ-ridden forms of privy was that of the water closet which was first patented in 1775 by a London watchmaker Alexander Cumming. This was a valve closet with an overhead cistern; by pulling up a handle the valve opened to empty the contents of the pan into a waste pipe and water entered to flush the pan. The basis of all subsequent W.C.s, its weakness was the sliding valve which was inefficient, Cumming, though, was not the inventor of the W.C. This was a godson of Queen Elizabeth I, Sir John Harington, who described his invention 179 years earlier, one which was not taken up due to lack of good water supply and drainage systems.

It was Joseph Bramah (page 75) who improved Cumming's water closet by redesigning the valve in 1778, so producing an appliance which sold in large numbers until overtaken 100 years later by the modern washdown system. Despite the great improvement which Bramah's closet represented, only comparatively few houses were fitted with valve closets and more common, at least until the 1890s, were the cheaper pan closet or long hopper closet, both of which were inadequately flushed so became very soiled. Over the years foul smells and gases permeated the house from such noisesome appliances.

Even more hazardous to health were the sewage disposal methods before the 1860s and for many years the introduction of the W.C. made this situation worse. Closets were tucked into corners of rooms, even into cupboards with almost no ventilation. The waste pipe was then emptied into a cesspool as before, but the connecting pipe provided a magnificent entry access into the house for foul gases. Safety measures came only slowly; in 1782 a stink trap was invented to keep out undesirable odours and by the 1840s cesspools were forbidden by law and the W.C. had to be emptied into the sewers, though even the town sewage system left much to be desired until later in the century (page 96).

A widely used alternative in later Victorian homes was the mechanical earth closet invented in 1860 by the Rev. Henry Moule, Vicar of Fordington (Dorset) (167). This was a wooden box with seat containing a bucket; above at the back was a hopper containing dried earth or ashes which, when a handle was pulled, would cover

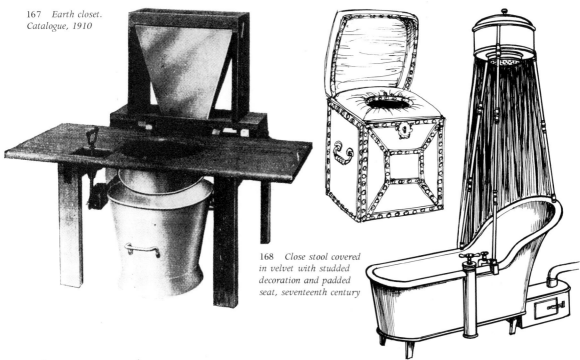

167 *Earth closet. Catalogue, 1910*

168 *Close stool covered in velvet with studded decoration and padded seat, seventeenth century*

169 *Bath and shower with curtains and hand pump for water, c.1855*

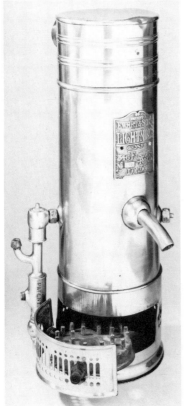

170 *Lighting gas geyser (Ewart), c.1920*

171 *Maughan's original gas geyser, 1868*

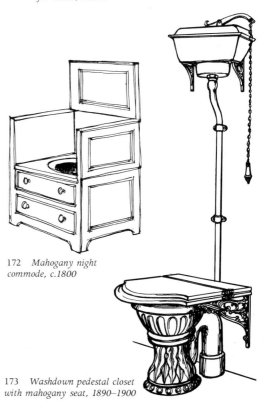

172 *Mahogany night commode, c.1800*

173 *Washdown pedestal closet with mahogany seat, 1890–1900*

the contents of the bucket. These closets were still on sale in London in 1910, price £1.50. Portable versions were used in bedrooms.

In the 1870s new designs of water closets were invented which gradually replaced the 100-year old Bramah type which had needed a valve (which became dirty and malodorous) to retain the water in the pan. Mr Twyford produced his Washout Closet in which water was always present in the pan, but a strong force of water was needed to flush. There was also a syphonic closet with two flushes but it was the washdown system which finally triumphed and is still in use (173).

FURNITURE

Technology brought change to the making of furniture chiefly with the sixteenth century introduction of panel and frame construction, with the development of the craft of veneering in the seventeenth century and with the progress of mechanization and introduction of new materials and methods from the later nineteenth century onwards.

The making of joined framed panelling utilized in wainscoting on walls (page 111) was also extended to oak furniture from the later fifteenth century. It could be seen in chests for storage, sideboard tables, buffets, settles and even cradles. The framework of horizontal rails and vertical stiles was jointed by mortise and tenon* then secured by oak pegs (175).

From about 1660 the 'age of walnut' in furniture-making began to replace the earlier 'age of oak'. Oak furniture was made of solid wood and decorated by painting, carving and inlay†. Walnut was particularly known for its fine veneers. During the last 40 years of the seventeenth century many European craftsmen, skilled in the latest constructional and decorative techniques of furniture-making, came to work in England; these included cabinet-makers from Flanders and, after the revocation of the Edict of Nantes in 1685, Huguenots also.

---

*The mortise is a cavity cut in one piece of wood to receive the tenon (or tongue) of the intersecting piece (175).
† A craft practised especially from the mid-16th century onwards in which coloured woods were sunk into the surface of a piece of furniture up to a depth of $\frac{1}{4}$ inch to give a pattern.

Veneered furniture is made by gluing thin sheets of walnut (sawn to 1/16th inch thick) to a carcase of oak or other wood. The carcase is a complete piece to which the veneer is added purely to give a decorative finish. Veneering is a highly skilled craft and, by cutting different parts of a tree, specific patterns may be obtained. For example, oyster designs come from small branches cut transversely, burr veneers from malformations in the tree. Though walnut was especially used, other woods, notably olive, laburnum, kingwood and maple, also give beautiful veneers. Marquetry was a further development of veneering. Here different coloured woods were cut into thin layers and made up into a design (180). This differs from inlay, which penetrates more deeply and in which each individual piece of wood is inserted separately; in marquetry the whole pattern is made up into a veneer and then glued on as a sheet. Marquetry designs were delicate, floral patterns predominating until about 1690, after which sea-weed versions in arabesques were fashionable. Parquetry is a further variation, being made into geometrical patterns.

Two other decorative means were introduced into English furniture-making in these years: japanning and gesso. Japanning stemmed from the Oriental lacquered furniture which was being imported and became very fashionable. The sap of a sumac tree which was used was not available in Europe, so an inferior imitation method was evolved called japanning. In this a coating of whiting and size was applied to the furniture and then several coats of coloured varnish onto which the design was painted in gilt. Relief patterns were built up with a paste made from gum arabic and whiting which was then gilded (178). Gesso (the Italian word for gypsum) furniture was particularly fashionable in England in the years 1690–1730. A paste made from whiting mixed with size was applied in successive coats to the piece until it was thick enough to carve or incise a pattern which was then gilded (179).

Papier mâché was a material particularly beloved of Victorian furniture-makers, though it had been introduced into England from France (hence its name) as long ago as the 1670s, and 100 years later Henry Clay took out a patent for

174   *Panelled oak chest inlaid with holly and bog oak, c.1610*

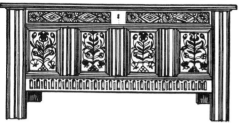

177   *Papier mâché tea poy. Decoration of gilding and mother-of-pearl, c.1850*

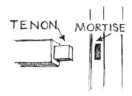

TENON   MORTISE

175   *Mortise and tenon joint*

176   *Dovetail key joint*

178   *Japanned cabinet on carved and silver stand. English, seventeeth century*

179   *Carved and gilded gesso on wood chest, c.1720*

180   *Cabinet-on-stand. Burr walnut veneer with panels of floral marquetry. English, late seventeeth century*

making it by pasting sheets of paper into moulds and baking them to produce a very hard substance. In the nineteeth century this method was still used, as was a different one where pulped paper was pressed between dies. In both instances many coats of varnish were applied so that the resulting hard material could be turned, planed or filed. Papier mâché furniture was then painted and gilded with ornate designs and finally varnished once more to give a brilliant 'lacquered' finish (177).

It was in the nineteenth century that the spiral spring was introduced into upholstered furniture. Before this horsehair or feather stuffing had been used, but the Victorian desire for comfort in beds and chairs brought about much thicker (if less elegant) upholstery which, after 1850, was more often sprung. Various types of spring had been in use during the eighteenth century, but for carriages, not furniture. Samuel Pratt patented a spiral spring in 1826 in order to make a swinging seat to be used on board ship to ease sea-sickness, but two years later he was envisaging its application to beds and cushions. The woven wire mattress supported on spiral springs was patented in 1865.

Although machines to carry out many woodworking processes had been invented by 1800, the furniture trade was slow to adopt mechanical methods. Then the unprecedented increase in population triggered off a heavy demand for new furniture and by the 1870s the process of mechanization accelerated and designs were being made specifically for such production means. Machines suited to furniture-making, powered first by steam and later by electricity, were gradually made available to saw, plane, bore, groove, carve, mould and mortise.

In the twentieth century all stages of manufacture can now be carried out by machinery of a piece based on the work of a talented furniture designer, so providing good furniture for sale at reasonable prices. At the same time mechanization has not completely replaced handcraft, but a hand-made piece will be very expensive. Much of the best furniture now available is the product of both methods. All modern technological means of reducing costs and speeding production are employed, but the use of good materials and careful hand-finishing ensures a quality article.

While machinery has modernized the processes of furniture-making, the introduction of new materials and different methods of handling existing ones have been developed during the twentieth century, particularly since 1950. One example is metal. Japanned iron tubing had been widely employed during the nineteenth century for making bedsteads, but the advent of tubular steel (later chromium-plated) led to its more extensive use, for instance for the cantilevered chair first introduced by Marcel Breuer in 1925, and its reappearance in a variety of guises since. For lightness steel is often replaced by aluminium (182).

A second instance, and one more generally employed in domestic furniture, is lamination. The new bonding resins developed for needs during the Second World War have made the extensive moulding and bending of plywoods by electrical means possible without danger of damage and splitting (183). These curved shapes are especially suited to seating furniture. The production of a range of plastics since the 1950s (page 83) has introduced a new tough, inexpensive material to furniture-making. In particular, polypropylene moulded by injection methods has made possible all kinds of unbreakable, stain-proof furniture especially useful in kitchens, bathrooms and nurseries (181). In a different form, foam plastics have become indispensable for upholstery needs.

181 *Hille chairs with shells moulded from polypropylene covered with a 'Cambrelle' non-woven nylon fabric which becomes an integral part of the shells during the injection moulding process*

183 *Bending lengths of beechwood under pressure for use in Windsor chair backs. A critical process which must avoid splitting or crushing of the wood*

182 *Cantilevered tubular steel dining chair, 1936*

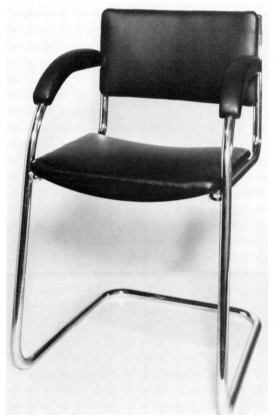

CHAPTER SEVEN

Textiles

NATURAL FIBRES

The making of all the principal fabrics from natural fibres – cotton, linen, wool and silk – dates from ancient times. In Europe the mechanization of these processes, led by Britain, spearheaded the Industrial Revolution. Cotton was in the vanguard.

Cotton is grown in over 60 countries. The raw material is the soft, white, fibrous substance which envelops the seeds of the cotton plant (gossypium) and, until the eighteenth century, was imported into Europe chiefly from the Middle East and India. During the eighteenth century the West Indies was Britain's main source of supply, after which it came chiefly from the Southern States of America. European trading of negroes to the New World had been in operation since the early sixteenth century and by 1700 the so-called 'triangular trade' was well established. The British had a large part in it and Bristol merchants sailed to Africa to exchange their manufactured goods for negro labour which they then took to the West Indies and America, returning to Britain on the third leg of the journey with raw cotton received in lieu of the slaves. By 1800 the cotton trade in England had increased immensely. The port of Liverpool had replaced that of Bristol as its centre and the making of the textile was gradually established in the suitably damp climate of Lancashire.

Cotton was the first major industry to become mechanized, to adopt factory methods of manufacture and handling of labour and to power its machines, first by water then by steam. Production increased very fast to supply not only a rapidly expanding population but also an extending empire (page 15).

Linen was woven from flax in the Middle East as early as 3000BC, with Egypt as the chief producer. Flax fibres are strong and have good spinning qualities. Traditionally the flax plants were pulled by hand, bound into bundles and dried. They were then combed to remove the seed capsules and soaked in water to separate the fibres from the hard centre. These fibres were then beaten and washed. Although the linen industry had been developed so early, it was difficult to mechanize, and so industrialization came late. The hackling process (where the fibres are combed to separate the longer 'line' ones from the shorter 'tow' ones) was a particular problem and the lack of elasticity in the spun flax caused breakages in power loom operation. It was the second half of the nineteenth century before these difficulties were satisfactorily overcome.

In Europe wool is obtained from sheep, an animal domesticated in Britain before the coming of the Romans. From the early Middle Ages until the rapid development of the cotton industry in the later eighteenth century, British woollen cloth was world famous and the country's greatest revenue earner. Cross-breeding of sheep has been experimented with over the centuries to produce finer, warmer and more durable wool. In the eighteenth century the natural English stock was cross-bred with the Spanish merino sheep whose descendant is the Southdown. Certain fleeces are suited to specific needs, for example Welsh wool for flannels, Shetland for knitting yarns and Cheviot for tweeds (this is said to have been perfected by a cross between the indigenous Cheviot sheep and some merinos carried on the Armada ships in 1588). Hand shearing of sheep was gradually replaced in the late nineteenth century by the development of machine shearing in Australia.

Silk is the finest and most sought after of all natural fibres and the only one to be available in a continuous filament. In the ancient world the best quality filament was obtained from the cocoon of the silk moth (*bombyx mori*). The grubs of the moth feed on mulberry leaves and they spin their cocoons, at the appropriate time in their life cycle, by exuding a viscous substance in a fine filament, one from each of two glands on either side of the body. The larva fastens one end of the filament to a twig and spins a cocoon, composed of continuous threads 800–1,200 yards long, round itself. In the natural process the moth pushes the threads aside when it is ready to emerge. Under sericulture the chrysalis is killed by suffocation with steam or hot air and the continuous thread unravelled.

The Chinese initiated silk cultivation in very remote times, some sources say as long as 8,000 years ago. They guarded their secret closely but, eventually, the knowledge was acquired by India (about 1000BC), then Japan (third century AD), Byzantium (sixth century) and later came to Europe via North Africa and Asia Minor. Italy was making silk in the early Middle Ages and France was weaving it by 1500. It was the seventeenth-century Huguenots who established a silk industry at Spitalfields in London.

TEXTILE PROCESSES

A series of preliminary operations are necessary to prepare the natural fibres of cotton, linen and wool for spinning. In all cases the raw material has to be cleaned and extraneous matter removed. Wool from the shearer has to be graded then washed. Flax has to be combed, beaten, then hackled (page 132) and cotton beaten to break up the pods and the seeds removed. Until the eighteenth century all the processes were per-

formed by hand, but by the 1780s the need to increase production to supply a larger population combined with pressure from handworkers for higher wages spurred on attempts to make machines which would carry out these tasks.

As the cotton trade between the New World and Britain built up, several ways of mechanizing the cleaning of the American raw cotton were devised to enable the plantation owners to speed up transportation and so reduce costs. Batting machines were designed which opened up the pods and disposed of the seeds. In principle a machine consisted of a rotating cylindrical cage of cane which retained the cotton but allowed the seeds and fragments of pod to fall through a mesh. The seedless cotton was then compressed into bales for shipment. More advanced machines were then devised, for example, the scutching* and blowing machine of the 1780s which broke up the bolls as the raw cotton was passed between the rollers to be beaten up by metal bars. The released seeds fell through a grid and finally a current of air was blown through the cotton to remove dirt particles. A further advance was Eli Whitney's cotton gin (engine) of 1794, improved in 1796 as a saw gin which was horse-driven, so enabling one man in charge of the operation to replace 50 men who had worked the roller gins.

Meanwhile in Britain a roller mechanism for scutching flax, powered by water, was devised in the 1720s, but more than 100 years elapsed before a satisfactory hackling machine (page 132) was adopted and, even in 1850, flax for fine linen was still hand-hackled (186).

The woollen industry received a great boost after the invention in 1801 (greatly improved in the 1830s) of a rag-tearing machine called a 'devil', which provided an enormous source of cheap raw material. Wool rags were fed by rollers into a drum fitted with metal teeth which tore up the material as it was rapidly revolved. The bulk of this product was then spun (with the addition of some new wool) into yarn called shoddy. Yarn made from better quality rags was termed mungo. The residue not suitable for spinning was sold as agricultural fertilizer and, by the 1850s,

---

* Scutching is the process of dressing a fibrous material such as cotton or flax by beating.

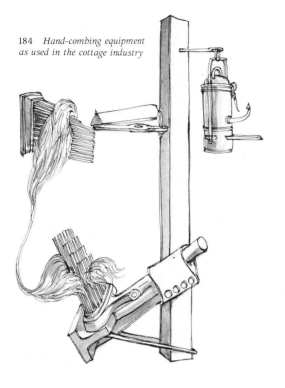

184   *Hand-combing equipment as used in the cottage industry*

185   *Dr Edmund Cartwright*

186   *Wooden comb for hackling flax, fourth century* AD

187   *Wooden hand card*

188   *Machine for combing cotton, 1894. An improved form of Heilmann's wool comber of 1846*

for use in manufacturing flock wallpapers (page 111). By the late nineteenth century the making of shoddy had grown to such importance that 40 per cent of the woollen industry's raw material came from this source.

After cleansing the most important preliminary processes before spinning are carding and combing. The purpose of carding is to disentangle and straighten the fibres. Before mechanization this was done with a pair of hand cards made of wood faced with leather and fitted with small wire hooks (187). Machines for carding cotton were devised in the 1740s and greatly improved by the 1770s when Sir Richard Arkwright (1732–92) (190) made his version. In this, cotton was fed by fluted rollers on to a fast-moving large cylinder and was then carded between this and a series of flats which covered it (189). By the end of the eighteenth century two main designs of carding engine were available: one using flat cards (found to be most suitable for cotton) and a roller and clearer machine in which pairs of rollers over the cylinder carried out the carding in conjunction with it (this was more suited for use with wool). During the nineteenth century power-driven batting machines were developed which fed the raw cotton directly on to the carding engines which have continued to be improved upon and speeded up into modern times.

Combing, using hand combs with rows of long metal teeth, penetrated deeper into the mass of fibres than carding, making them lie parallel. With wool it was an especially important process as it separated the long fibres from the shorter ones, the longer ones being suited to spinning into worsted yarns and the shorter into twisted woollen yarns. Hand-combing was hard, skilled work and, in the second half of the eighteenth century combers repeatedly struck for higher rates of pay (184). Under pressure, employers sought a machine to replace their rebellious workers. Edmund Cartwright (1743–1823) (185) patented his wool-combing machine in 1792, but its products were inferior to hand-combing. Other inventions followed, also further industrial unrest. The first successful machine in Britain, based upon French ideas, was Platt's design of 1827, which continued in use until the Lister-Donisthorpe machine of 1851 revolutionized wool-combing in Bradford, replacing hand-combing, but James Noble's machine of 1853 became the standard type. The improved version of 1862 operated with one large horizontal circular comb with two smaller ones revolving inside. The wood was fed on to the larger comber and the small combs gradually took the wool from it (188).

One other process was found beneficial after carding cotton: this was drawing which drew out the slivers and evened out their inequalities, so, after spinning, producing a yarn of greater evenness and regularity. A drawing frame was developed into the late 1770s by Sir Richard Arkwright for this purpose (191).

Spinning

The process of spinning is to draw out the fibres and twist them into a thread. There are three actions in this operation: first to attenuate the raw material, second to twist the drawn-out fibres and lastly to wind the yarn on to a holder. The traditional method from antiquity until the Middle Ages was by spindle and whorl (193). In this the unspun fibre was carried on a stick (the distaff) held in one hand by the spinster who drew out with the other hand fibres which she teased between finger and thumb into a thread which was attached to a weighted stick (the spindle). This was notched at one end to hold the thread and at the other was fitted with a disc of wood or pottery (the whorl) which was set rotating to twist the thread.

The spinning wheel (most commonly called the great wheel) greatly speeded up this process and made it more efficient (192). Introduced from the East, it was in use in Europe from the fourteenth century. The spinster stood, turned the wheel with her right hand and drew out the fibres with her left. The spindle, mounted horizontally on bearings, was rotated by means of a driving band from the wheel. The two operations of spinning the thread, then winding it on to the spindle, had to be done separately and consecutively. The introduction of the flyer and bobbin about 1480, which enabled the two processes to be carried on continuously without the spinster having to keep stopping between operations, was a **great**

189   *Carding engine of the type designed by Arkwright, c.1775*

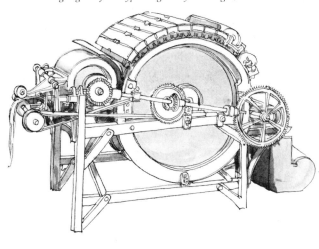

190   *Sir Richard Arkwright*

191   *Arkwright's drawing frame*

192   *The Great Wheel*

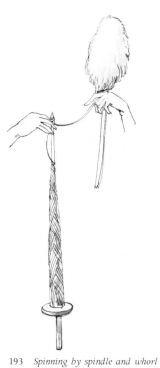

193   *Spinning by spindle and whorl*

194 *Cottage spinning and reeling of linen yarn, eighteenth century. Two women are spinning with Saxony wheels, the third is using the type of wrap reel invented by Arkwright for winding yarn into hanks of measured length*

195 *Interior of a mule spinning factory, 1835*

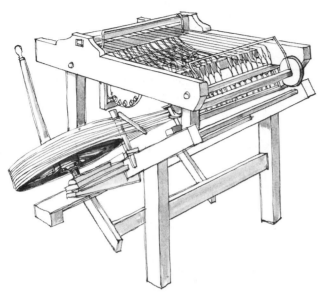

196   *James Hargreaves' spinning jenny*

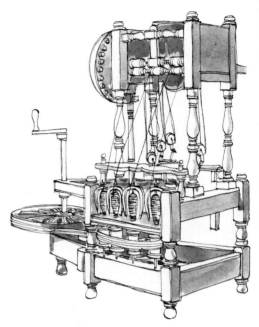

197   *Spinning frame (later, water frame).
Sir Richard Arkwright, 1769*

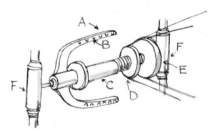

198   *Jurgens' flyer and bobbin as used in
the Saxony Wheel, see fig. 194
A Flyer B Hooks C Bobbin D Bobbin pulley
E Flyer pulley F Supports*

199   *Ring spinning frame, 1926. Portion of ring frame
shown here, in full length 400 spindles would be
carried. In this model one side was converted in 1961 to
the Casablancas High Draft System which enables the
yarn to be spun directly from slubbing, thus dispensing
with intermediate and roving frames*

advance and the incorporation of a treadle mechanism in the sixteenth century was of further assistance. These two additions were included in the Saxony Wheel designed by Jurgens of Germany, which remained the standard pattern until the eighteenth century (194, 198).

By 1760 mechanization of weaving was causing hold-ups as hand-spinners could not produce sufficient yarn to supply the looms. The first successful 'spinning machine' devised to solve this problem was James Hargreaves' (d.1778) spinning jenny* of 1764. This had eight vertical spindles operated on the great wheel principle, with eight driving bands enabling eight threads to be spun simultaneously (196). By 1766 Hargreaves improved his design to take 16 spindles and set the wheel vertically, so that it was easier for the spinster to handle. The jenny was patented in 1770 and, despite violently destructive acts by angry hand-spinners, was widely used, especially in Lancashire, for spinning cotton yarns.

The weakness of the jenny was that its yarn was not strong enough for warp threads. A machine which overcame this difficulty was Sir Richard Arkwright's (page 135) spinning frame of 1768, which made use of rollers to draw out the rovings before being passed to the spindle. Arkwright's frame could not be handled by a human operator and required more power. He designed it to be geared to a horse-mill, but it was soon adapted for water power (using 24 spindles) and became known as the water frame. It was then adapted to spin flax and wool as well as cotton (197).

In 1779 Samuel Crompton (1753–1827) devised a machine which became known as the spinning mule because it was a cross between Arkwright's method of drawing out the fibres by rollers and Hargreaves' movable reciprocal carriage, so containing the best features of both (211). The mule produced finer softer yarns than before. First hand-operated and made with 30 spindles it was soon enlarged, later taking up to 1,000 and operated by water power (page 132).

By the end of the eighteenth century spinning machines were being powered by steam (page 15)

and during the nineteenth several useful improvements were made (195). The throstle machine was designed in 1815 to produce a strong yarn suited for use in power looms. In 1830 Richard Roberts patented a shaper device (the cop) which made mules self-acting. At this time also the most important development of the century, the ring spinning frame, which was produced in the USA, was a great advance on both the throstle and the mule (199). It worked faster, wound and spun continuously and could be operated by unskilled labour. Much improved versions appeared in the 1880s and this remained the dominant machine during the twentieth century, though newer concepts were behind the break-, rotor- and twistless-spinning methods developed in the 1950s and 1960s. These made use of an air vortex or an electrostatic field to build up the fibres into a yarn then twist it.

Silk

The processes necessary to prepare silk for weaving are different from those for wool, cotton and flax because the silk is already in the form of a long filament extruded by the silkworm (really silk caterpillar) (200) and wound into a cocoon (page 133). These filaments have to be unwound –

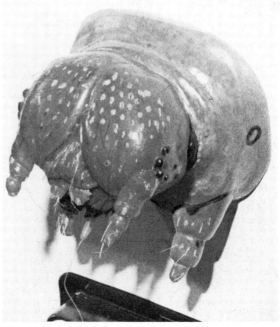

*Jenny, like 'gin' and 'ginny' is a colloquialism for 'engine'.

200 *Model of enlarged head of a silkworm*

a process called reeling which produces raw silk – and as they are too fine and delicate to be reeled individually, filaments from several cocoons are handled together and are washed to remove the gummy sericin. During the nineteenth century various means were perfected to beat the cocoons and so free the silk from portions of chrysalis and other extraneous matter. This made it possible to utilize a greater proportion of the silk than before. After reeling, the grouped filaments are twisted to make a continuous yarn of suitable thickness for weaving. This process, known as throwing, was originally carried out by hand, using a double wheel and cross (201).

In Britain mechanization of these operations was not begun until the eighteenth century, though in silk-producing parts of the world this had been done long before. In China a reeling machine was being operated in the first century BC and in Europe the Italians were throwing silk in the thirteenth century by means of water-powered mills. In Britain in 1718 Sir Thomas Lombe (1685–1739) patented a silk-throwing machine based on the Italian design and this, and

others like it, continued in use well into the nineteenth century. Similarly, John Heathcote (1783–1861), who had already patented his lace-making machine in 1808, invented a machine for reeling cocoons; it was many years too before this was superseded.

More than half the silk contained in the cocoons was not reelable, either from damage to the cocoon or because the filament would not unwind. This was known as waste silk, but after washing much of it could be hand-carded and -combed, then spun on a wheel. During the eighteenth century in Britain spun silk became an established cottage industry. It was then found that, if the filaments were cut up into short lengths, they could be machine spun. Mechanization began in 1792 in Lancashire. Machines cut up the filaments into lengths of one or two inches, which corresponded to cotton fibres, then cotton spinning machinery could be adapted for the work. In the 1830s mechanization of the spinning of longer fibres was devised and this gradually ousted the earlier method.

In the second half of the nineteenth century

201 *Silk twister's wheel, cross and ruler (throwing process)*

considerable advances were made in designing machines to handle the longer silk filaments. These improved the combing and dressing of the silk, softening it and giving greater lustre and quality. From the 1880s wild silks also were spun and woven. These were silks from cocoons of wild silkworms which fed on a number of different trees. The tussah (tussore) moth of India is one well-known example.

Warping and weaving

All woven fabrics consist of two sets of threads crossing one another at right angles. One set, the warp, is fixed to the loom; the other, the weft, is interlaced across the warp by means of a shuttle. Warp threads need considerable preparation: the correct number of threads (6,000 or more) must be cut to an exact length (probably about 100 yards) and they must be laid out untangled and parallel. Warping is the name for this preparation which traditionally was done on a warping board with pegs to measure the lengths. This was superseded in the seventeenth century by a warping mill, turned by hand or powered by horse or water, where single threads from bobbins were converted to a rope of untwisted yarn (202). Mechanical methods were perfected in the 1870s. After warping the threads were chained (plaited) to prevent re-tangling and were then sized to offset friction wear during weaving (203, 212). These operations were also mechanized later.

The warp threads were then wound on to the warp beam and drawn-in, that is, threaded through the heddles and reed and tied to the cloth beam on the other side of the loom. (Mechanization of the drawing-in process came late, not until the 1880s). The heddles were raised and lowered by the weaver's feet on the treadles which operated pulleys known as 'horses'. As one heddle was raised, selected warp threads rose with it and left a space (the shed) through which the shuttle, containing the weft thread wound on to a bobbin, could pass and so produce a line of weaving (204).

For centuries the shuttle was laboriously passed (an operation called throwing) by hand from side to side of the loom and, unless the weaving was of narrow width, two people were needed for this, one on each side. John Kay's (1704–70) (208) invention of the flying shuttle in 1733 was a great advance (209). In this the shuttle was thrown across the loom along a defined path by a mechanism operated by the weaver: a faster and more accurate process (210). His son Robert invented a multiple shuttle-box in 1760 which made the system work with several shuttles loaded with different coloured weft yarns.

Attempts to design power looms had been made since the sixteenth century and powered narrow gauge looms were made by 1600 to produce tapes and ribbons. It was Dr Edmund Cartwright (1743–1823), a Leicestershire rural rector, who finally succeeded in designing and patenting a practical wider power loom. After discussion with Manchester cotton manufacturers in 1786 he soon produced improved versions which could be powered by horses, water or steam, but, partly due to the great industrial unrest caused by its threat to employment, the power loom was slow to replace hand-loom weaving and it was only between 1850 and 1890 that steam-powered weaving eventually triumphed in all the main fabrics (205).

Patterned weaving was slow to develop as the problems in devising a mechanism to achieve a figured material were considerable. The punched card system was largely an eighteenth-century French development which culminated in the apparatus introduced in 1801 by Joseph Marie Jacquard (1752–1834) for use in the silk industry in France (207). The pattern to be woven was drawn on squared paper and the punched cards were prepared for this. The warp threads were then forced, mechanically, to follow this pattern according to the holes punched in the cards. The Jacquard attachment, which could be used with hand and power looms, was not adopted extensively in England until it was adapted in the 1820s for cottage and factory use.

Nineteenth-century improvements to the power loom included an automatic stopping device if the weft supply failed and a warp-protector motion. The most important advance was the invention in 1895 by the English-born American James H. Northrop of automatic means to replenish the weft in the shuttle. His Northrop

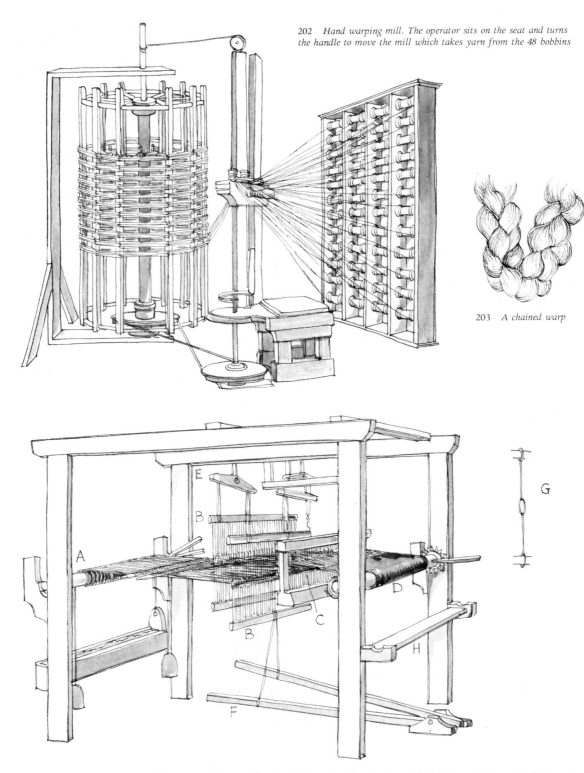

202   *Hand warping mill. The operator sits on the seat and turns the handle to move the mill which takes yarn from the 48 bobbins*

203   *A chained warp*

204   *Typical hand loom, A Warp beam or roller B Heddles (healds) C Reed enclosed in batten D Cloth beam E Horse F Pedal G Leash H Weaver's seat*

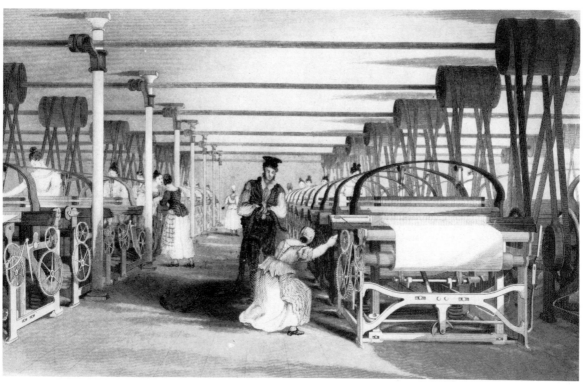

205   *Power loom weaving, 1835*

206   *Northrop automatic loom, 1939. Whenever the shuttle runs short of weft, a full bobbin from the rotary battery is automatically inserted. A feeler passes through a hole in the left-hand shuttle box when the shuttle is momentarily stationary between picks and, if it finds the bobbin nearly empty, actuates the changeover mechanism which operates when the shuttle next reaches the right-hand box. The loom is also equipped with a mechanical warp stop motion. One weaver can look after as many as 48 of these looms*

207   Joseph Marie Jacquard

208   John Kay

209   Flying shuttle

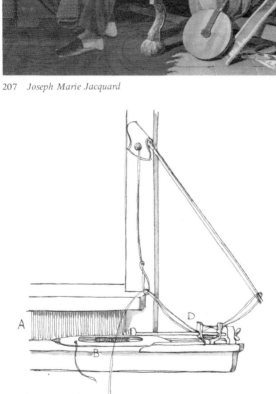

210   John Kay's flying shuttle A Reed B Shuttle C Picking Stick
D Picker and picker return spring

211   Samuel Crompton

loom became synonymous with automatic loom. In it a full bobbin was inserted when required, a thread cutter cut the weft of the nearly empty bobbin and, as it was ejected from the shuttle, the loom action resumed with the new one (206).

The modern loom works more quietly, it is powered by electricity and the shuttle can be propelled by a jet of air or water. Alternatively, shuttleless looms have been developed, using the napier or the gripper system.

Finishing processes

After a textile has been woven there are still a number of processes which it must undergo before it can be used. The material is washed after it is taken off the loom. Cottons and linens then had to be bleached to improve their colour and texture. Traditional methods were to bake the fabric in the hot sun or boil it in alkalized water known as lye* then steep it in buttermilk. The treatments were alternated and took a very long time. In the eighteenth century sulphuric acid began to replace the milk and by its close the use of chlorine had radically improved the process (page 77). After the Second World War new bleaching agents set a much higher standard of 'whiteness'.

A polish or glaze was then imparted to cottons and linens. Wood and metal hand glazers and polishers were used for centuries as were screw-presses (215, 218). Alternatively, or additionally, the fabric was calendered by a heavy roller (calender) to exert pressure, often combined with dry heat or steam. The process of mercerizing, which gives a permanent gloss and lustre to cotton, was named after the English calico printer and dyer John Mercer (1791–1866), who experimented with the use of caustic soda on the fabric and patented his discoveries in 1850. A satisfactory result was not achieved until it was found by H.A. Lowe in 1890 that it was mercerizing under tension which imparted the permanent gloss as well as strengthening the cotton to take the dyes better. An efficient process was finally developed in 1895.

Since 1950 a variety of finishing processes have evolved for different fabrics to make them flame-resistant and water-repellant, also both crease-resistant and permanently pleated. These durable press effects are achieved by impregnating the material with a cross-linking agent, then pressing it under heat to give permanence.

Woollen cloth needed to be fulled, teazled and sheared. Fulling cleanses and thickens the yarn; it was particularly suited to felts and heavy woollens. This process (216), in which the cloth was soaked in water and fuller's earth and beaten with hammers, was one of the earliest to be mechanized, the fulling mill being powered by a water wheel as early as the twelfth century (page 24).

After the fulled cloth had been washed and dried it was teazled to raise the nap. This was done originally by a hand-held teazle cross, a wood frame containing thistle heads (213). Nap-raising teazle machines, known as gig-mills, were introduced in the fifteenth century, though outlawed in England for about 100 years until the 1640s. In this teazles (natural or wire hooks) were fixed into a cylinder which was rotated by horses or water wheel and later by steam power. After the nap had been raised the cloth was then cropped to give an even, smooth pile. For centuries this was done by hand, using enormous cropping shears about four feet long (214). Powered machines to do this were devised from the late eighteenth century.

Dyeing

Through the centuries the aim of the ancient art of dyeing has been to produce a fabric colouring which is impervious to washing, rubbing and exposure to sunlight. Natural plant, shell and insect substances were used by early societies, for example, woad (blue), saffron (yellow) and madder (red) from plants, the mollusc Purpura (purple) and cochineal and lac (reds) from insects. Empirical methods disclosed that some mineral substances were fast to light; these included the iron earths (reds, yellows, browns), lime, gypsum and clay (white and cream) and soot and coal (greys and blacks). The use of solvents such as urine, sea-water and saliva gave more permanent results, likewise lichens dissolved in water.

All these processes produced only temporary colouring, but it was found that if an auxiliary

---

*Most commonly made by pouring water through wood ashes or adding urine to the water.

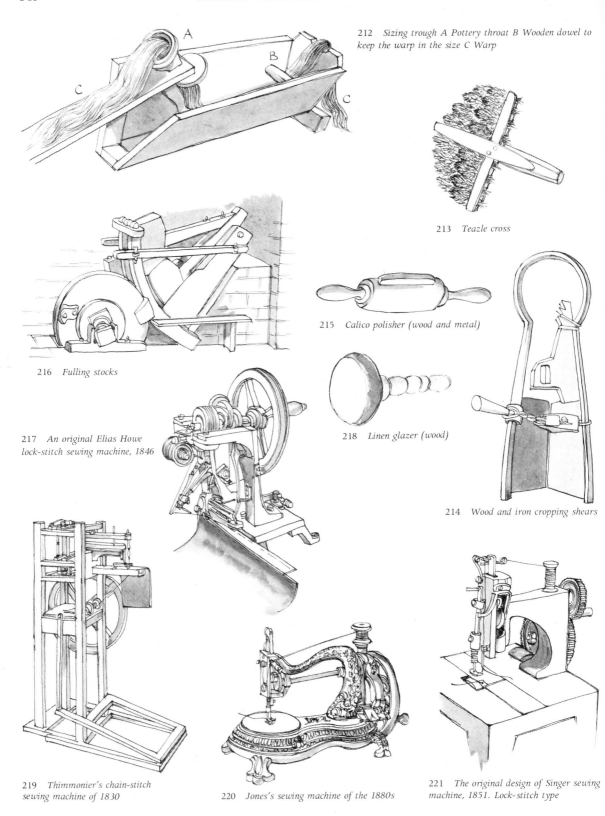

212   *Sizing trough A Pottery throat B Wooden dowel to keep the warp in the size C Warp*

213   *Teazle cross*

215   *Calico polisher (wood and metal)*

216   *Fulling stocks*

217   *An original Elias Howe lock-stitch sewing machine, 1846*

218   *Linen glazer (wood)*

214   *Wood and iron cropping shears*

219   *Thimmonier's chain-stitch sewing machine of 1830*

220   *Jones's sewing machine of the 1880s*

221   *The original design of Singer sewing machine, 1851. Lock-stitch type*

chemical was added to the natural dye substance, this would unite the dye with the fabric by better absorption, so giving a more lasting colour. The chemical needed varied with the dye substance used and trial and error decided which chemical gave the best results. Such chemicals are called mordants, from the French word for 'to bite'.

Research and development in the chemical industry in the seventeenth and eighteenth centuries led to the use of a number of different mordants (page 78), tin and lead, for instance, and many attempts were made to create synthetic dyes. Success was not achieved in this endeavour until Sir William Henry Perkin (1838–1907) made the first aniline dye in 1856. The chemical base aniline had first been obtained in 1841 by C.J. Fritzche, who distilled indigo with caustic potash; he named it aniline from the Sanskrit word for the indigo plant. Since then it has chiefly been made from coal-tar, hence aniline dyes are often referred to as coal-tar dyes. Perkin's first dye he called mauveine and European research soon produced several more – magenta, alizarin, fuchsia – all brilliant, but variably fugitive to light (88).

Research continued during the rest of the nineteenth century and the twentieth towards the achievement of dyes which were brilliant, varied and fast. Today, the scope of synthetic dyes is vast and the subject complex. Different dyes are suited to specific fabrics. The reactive dyes introduced in the 1950s by ICI marked a notable advance in that they form a chemical linkage with the fabric fibres, so remaining brilliant and fast, even with cold-dyeing methods, with washing, dry cleaning, wear and sunlight. Reactive dyes are most successful with cotton and cellulose materials, but more recent development has proved their suitability to both animal and synthetic fibres.

THE SEWING MACHINE

In 1790 an Englishman, Thomas Saint, designed and patented a sewing machine. He made drawings of it and wrote a description, but it seems that he did not actually make it and it was not until 1830 that the Frenchman Barthélemy Thimmonier invented the first satisfactorily functioning mechanism to sew (219). He was a tailor from the Lyons area and his single-thread chain-stitch machines were employed in making army uniforms. Despite the destruction of his factory and machines as well as physical violence perpetrated on himself by angry tailors, he devised improved machines and was soon back at work.

A machine sews in a different manner from a hand-sewer. The latter passes the needle and thread through the material and pierces it again from the other side. It is not easy to design a machine to imitate this process. In machine sewing the successful method, followed from the early designs onwards, has been to use an eye-pointed needle which will convey the thread through the cloth without having itself to pass through completely. There are three types of stitch: the single chain, the double chain and the lock stitch. Thimmonier's machine was of the first type, having a single thread held by a looper as the needle rises and falls so each stitch is secured by the one before. In the other two systems two threads are employed, but the lock stitch has proved the best method for domestic machines. In this there is a needle thread above the material and a bobbin thread under it and the two are locked together at each stitch by the upper thread being passed round the lower and tightening it in position.

The lock-stitch sewing machine was developed chiefly in America, first by Walter Hunt in 1832–4, then Elias Howe in 1846 (217) and, probably best known, is Isaac Merrit Singer's design of 1851 (221). This machine appeared on the market at the opportune moment to handle the yards of sewing required in the fashion for crinoline skirts, the circumference of which reached its maximum in 1859–60.

Since the 1850s many improvements have been made (220) such as the introduction of positive take-up in 1872, the reversible feed of 1919 and the increasing availability of varied and decorative stitches. The first domestic electric sewing machine was marketed by the Singer Co. in 1889, though these were not in general use until after the 1920s. Domestic zig-zag machines which revolutionised home sewing came from Italy (Necchi) in 1947 and, recently, the last word in sewing machines is the Singer electronic model controlled by a microprocessor (page 171).

SYNTHETIC FABRICS

The greatest textile event in modern times has been the emergence and meteoric growth of the synthetic fibres industry. These man-made textiles are, like natural ones, derived from living organisms but, unlike those which came from still-living plants, insects or animals, they are produced from those which millions of years ago died and were slowly transformed into minerals. Synthetic fibres are made by the chemical treatment of materials such as wood pulp or, now more commonly, derive from extracts of petroleum and coal. The substances are transformed to a viscous liquid which can be extruded through a series of fine holes to form filaments which may be twisted, woven or knitted into fabrics. They may also be cut into short staples, then combed, drawn and spun as in the natural fibre processes; they may also be blended with such natural fibres before spinning.

It was the English scientist Robert Hooke (page 13) who put forward the idea in his *Micrographia* of 1664 that such a fibre could be made and extruded in a similar manner to that demonstrated by the silkworm. During the following 200 years others theorized and experimented with the idea: the scientist Réamur in France in 1734, the silk weaver Schwabe in England in 1842, the Swiss Audemars in 1855 and even Sir Joseph Swan in his research for carbon filament electric lamps in 1883 (page 119).

It was known that cellulose is the chief component of substances such as cotton, wood and paper and, after years of experimentation, the Frenchman Comte Hilaire de Chardonnet (1839–1924) produced his first fabric from a nitrocellulose solution which was extruded through fine holes to give spinnable filaments. He exhibited articles made from the fabric in the Paris Exposition of 1889 and two years later began commercial production of his 'artificial silk' at Besançon.

However, nitrocellulose made from cotton and nitric acid had a tendency to explosive qualities, so safer alternative ways of making this artificial silk were sought. The cuprammonium process was developed in Germany and commercial manufacture began there in 1899. More successful was the viscose process resulting from the researches of a group of chemists in the 1890s. The British rights for this process were bought by Courtaulds in 1904, who have since carried out much of the research and development. The fourth method, the acetate process (page 85) was undertaken just before the First World War and its commercialization achieved by the Dreyfus brothers who marketed the product under the trade name of 'Celanese' in 1921. It was during the inter-war years that artificial or 'art' silk enjoyed its boom period. America adopted the name rayon in 1924 to designate the fabric.

With the advent of nylon a new synthetic fibre industry was created. It was the first completely synthetic fibre, produced entirely from mineral sources. A polyamide fibre, it is composed of nitrogen and oxygen (derived from air) and hydrocarbons (derived at first from coal but now from oil and natural gas). Nylon was developed in the Du Pont laboratories between 1927 and 1938 at a cost of over 27 million dollars. It was made into stockings in 1939 and soon into other articles. It describes a family of synthetic fibres which were later made in many countries under different trade names: Enkalon, Perlon, Nomex, Banlon, etc. The British manufacturing rights were bought by ICI in 1940 and British Nylon Spinners (a joint ICI-Courtaulds Co.) were established to make Bri-Nylon in Britain.

It was during the Second World War that Whinfield and Dickinson in England made a new fibre from a polyester derived from a by-product of cracking of petroleum. During war-time Du Pont developed the fibre and marketed it as 'Dacron'. In 1955 ICI produced their fibre which they called 'Terylene': the bulked form of this was the ubiquitous 'Crimplene'.

At first traditional textile machinery was adapted to process man-made fibres, but soon it was realized that, in order to take effective advantage of their remarkable properties, new machinery would be needed. Various ways of spinning were developed: melt-spinning was used for nylon and polyester, dry-spinning for cellulose fibres such as tricel, wet-spinning for acrylic fibres, courtelle and acrilan. Processes were evolved for bulking synthetic yarns to give them some of the qualities possessed by staple yarns. Others were found to take advantage of the high elastic quality of the materials by developing them as drip-dry fabrics and permanently pleating them by heat and stretch processes.

CARPETS

Until about 1700 in England home-produced carpets were embroidered or of 'turkey-work', a name given to a type of carpet-knotting copied from Turkish methods. After 1685 many of the French Huguenot carpet weavers fled to England and set up their craft where, with royal patronage, it flourished in centres which became renowned for carpet production: Wilton, Axminster, Kidderminster.

It was Erastus B. Bigelow who devised the first steam-powered loom to weave carpets in Massachusetts in 1839; a few years later he adapted this for making the carpet with an uncut pile surface, familiarly known as a Brussels carpet. The loom was adopted in Britain in mid-century and a jacquard attachment was employed for patterned weaving. Soon the use of the loom was extended to the Wilton type of cut pile, the loops of the pile being raised by one set of wires while another set with a knife blade attachment was incorporated to cut the pile. A little later the power loom was making the cheaper 'tapestry' carpets and, finally, chenille carpets also.

Tufted carpets were introduced from the USA, based on the idea of candlewick bedspreads. Sewing machines were adapted to make tufted fabrics in the 1920s, but it was 1950 before the wide complex carpet tufting machine was designed. The tufts are held in position by an adhesive and the carpet is backed by polyurethane foam (page 85). After 1950 also the staple carpet fibre of centuries – wool – was gradually replaced by synthetic fibres.

CHAPTER EIGHT

Communication and Entertainment

LEISURE AND RECREATION

The revolution brought about by twentieth-century technology in the ways and means for passing the increased number of hours of free time has been every bit as far-reaching as that affecting all other areas of domestic life. One 'miracle' has succeeded another as, in less than a century, it has become possible in the privacy of one's own home to chat to a friend by telephone, to listen to a great orchestra playing on a recorded disc and to select from an almost round the clock series of programmes on radio. The greatest 'miracle' of all for many people has been the advent of television, the pictures of which can be beamed from a satellite far distant in space, vividly illustrating the happenings in the world outside at the very moment that they take place. And yet it has all come so quickly, so recently.

Before 1900 people made their own entertainment and, a century earlier, this was still severely limited in their homes in the winter months because the artificial lighting was so poor (page 115). People played cards, a wide variety of indoor games, made their own music from several different instruments, sang songs and dressed up to present home-made drama. Very popular was the projection lantern, later known as a magic lantern from its early seventeenth-century use in conjuring up pictures of ghosts and skeletons as if by magic. It was a wooden or metal box containing a light (a candle, oil, acetylene or later, electric lamp) which shone through a slide to produce a picture which, magnified by a projec-tion lens, appeared on a screen: the precursor of the modern projector for colour slides (222, 223). In 1814 Sir David Brewster adapted an earlier idea of using two or more inclined mirrors to produce multiple images of an object to make an instrument which he called a kaleidoscope. This simple means of being able to observe attractive symmetrical patterns in nature became popular, so much so that after it was patented in 1817, a million kaleidoscopes were sold in the first year of production.

Reading became very important in nineteenth-century homes, notably so when gas (and later electric) lighting relieved the eyestrain caused by the earlier low level of artificial illumination. News-letters (news sheets) had been available since the early seventeenth century and the first English newspaper had appeared in 1665. Books also had long been available, but were expensive. It was the introduction of mass production methods in the mechanization of the printing and paper-making processes (pages 152, 154) which sharply reduced book costs. Books of all kinds, as well as a wide variety of periodicals, were avidly read by the middle classes for entertainment and for information. When, after 1850, in many parts of Britain free municipal libraries began to lend books, working families began to read with enthusiasm also.

PAPER-MAKING

The word paper comes from the Latin *papyrus*, the paper-reed. This aquatic form of sedge, used

222   *Early magic lantern. From page 728 of Zahn's* Oculis Artificialis, *1702*

223   *Magic lantern by S.C. Hughes, c.1895*

extensively as a writing material in the ancient world and first made by the Egyptians some 5,000 years ago, resembled paper since it also derived from vegetable fibres. Papyrus was gradually replaced by parchment, made from animal (chiefly sheep and goat) skins. Parchment was a finer writing material than papyrus but, by its nature, was costly and it was the invention of letter-press printing (page 154) which hastened the replacement of parchment by the cheaper and more suitable paper.

Paper had been hand made by the Chinese as early as AD 105, using tree bark and textile waste. Five hundred years later the craft was practised in Japan and knowledge of it spread to the Arab world. The Moors then introduced paper-making into Spain where, by 1150, they had set up a mill at Jativa south of Valencia. From Spain paper-making spread in Europe, to Italy, France and Germany though, in England, paper was mainly imported until the 1590s.

Paper was made by hand until the nineteenth century, its raw materials chiefly linen and cotton rags. The paper-making process, whether by hand or modern machine, remains fundamentally the same. There are four stages of production. First the fibrous material must be prepared by making it into a pulp mixed with about 99 per cent water; then the paper is formed on the net; thirdly, the water has to be removed; and lastly the finishing operations are carried out.

In medieval paper-making in Europe the rags were beaten with the water into a pulp. As early as the twelfth century this process was hastened by the invention of the stamp-mill, which produced a pounding action as in a large-scale mortar and pestle. At first the mill was hand-operated, later water or wind power was introduced (225). A further advance was the Dutch invention in the 1670s of the hollander, a rotating cylinder on which knives were mounted to tear up the rags. For the second stage the pulp was transferred to a vat into which the paper-maker dipped his moulds. These were rectangular wooden frames to the bottom of which was fastened a row of stretched parallel wires. In 1500 there were about 28 wires to the inch. As the water drained through, the fibres felted and a sheet of paper was formed. Sheets were then removed, dried between layers of felt, squeezed in a press and, after being hung up to dry, were finished by sizing and being rubbed by a smooth stone.

The first machine to make paper continuously (about two feet wide and 40 feet long) was designed in 1797 by the Frenchman Nicholas-Louis Robert (224), but this was soon superseded by the more efficient and commercially successful machine of 1803 which the British engineer Bryan Donkin was commissioned by the Fourdrinier brothers to design to be an improvement on Robert's machine (226). The name of Fourdrinier is still given to the wet-end paper-forming section of the modern machine. Then, as now (despite numerous adaptations), the mechanism was designed to receive the prepared raw material, integrate the fibres to form a paper sheet, then dry, press and finish the product by a continuous succession of handling processes. During the twentieth century machines have been developed to deal with a tremendous range of papers of a much greater width and length and at a vastly increased speed; electric power has replaced other sources.

Since the 1850s demand for paper has continuously and dramatically increased, at first for books, newspapers, periodicals and letters and, more recently, to supply office and business needs. The raw materials available were not sufficient and led to an acute shortage of rags. These were supplemented by straw and, from 1855, the importation of esparto grass from Spain and North Africa. This was still not enough. The problem was solved by the use of wood pulp which provided a highly satisfactory raw material after chemical treatment of the wood chips by a soda or a sulphate solution, processes initiated in the 1850s.

Today much of Britain's paper derives from chemical and mechanical wood pulp*. In addition machinery has been developed to recycle waste paper by cleaning then repulping it; a process for

*Mechanical wood pulp, produced from finely ground logs with water, gives a paper of lower quality and less strength; it is used for newsprint and cheaper magazines and paperback books.

224 *Model of Robert of Didot's paper-making machine of 1797. Manually operated*

225 *Paper-making stamping mill, 1662. Powered by water-wheel. From* Theatrum Machinarum Novum *by G.A. Bockler*

226 *Model of Donkin's continuous paper-making machine. Early nineteenth century*

which Matthew Koops was granted a patent as early as 1800 but which, at the time, was not a success.

PRINTING

The invention of typographical printing in fifteenth-century Europe was an event of utmost importance affecting political, religious, cultural and social life. Before such printing was possible books had to be hand-written, so few duplicate copies existed and the (not infrequent) destruction of libraries by fire was an irreparable tragedy. In less than half a century more books had been printed than had been hand-written in the previous 1,000 years. Mystery surrounds this vital invention. Legends abound which credit the initial discoveries to, for example, Laurens Coster of Haarlem in Holland or to Fust or Gutenberg in Germany. It is also not known precisely what the first printing press was like or whether the European invention of movable type stemmed from the earlier ones in the Far East or was an independent development.

What is certain is that printing of pictures and words by means of carved stone or wood blocks preceded printing with movable type, also that, in the mid-fifteenth century, the means to carry out such printing had become available so that a method to bring these means together was being explored in several places simultaneously. There were four requirements: material to print upon, a formed shape to make the imprint, ink to print with and a press to transfer the ink to the material. Suitable paper had become available to print upon (page 152) and efforts were made to improve the early thin water-based inks which tended to smudge the print. It was found that lamp-black or charcoal ground in linseed oil provided a sufficiently high viscosity and printer's ink continued to be made in this way for over 400 years.

It also seems certain that the chief contributions towards casting pieces of movable type (made from an alloy of lead, tin and antimony), of making them accurately and standardized, of devising a means to hold the type in position for printing, of improving the quality of ink and of devising a suitable printing press were made, from 1448 onwards in Mainz in Germany by Johann Gutenberg (c.1399–1468). Soon printing presses were being set up in every country in Europe. William Caxton, who had learned his craft in Cologne, established the first press in England in 1476 at Westminster. After this the design of printing presses, which had originally been based on those in use for pressing linen or paper, was improved to be more suited to its specific task*, the handling of the type was speeded up and the type itself more standardized but, fundamentally, there was remarkably little change in the printing craft before 1800.

Increased demand in the nineteenth century for newspapers, books and a range of periodicals stimulated the mechanization of the printing industry in both presses and typesetting (227). First came the cast-iron press invented in 1800 by Lord Stanhope. This was still hand-operated, but its strength was a great improvement on wooden presses as was also its capability of printing the required large type-surface at one pull. However, it did little to increase output and for the mass production needed by the newspaper and magazine industry a powered press was wanted. The answer was provided by the steam-driven cylinder press designed in 1811 by the German Friedrich König which made history on 29 November 1814, when it was first used to print *The Times*. In the König press the type-bed was moved back and forward under a heavy inking cylinder, the hand-fed paper sheets then being pressed against the type by another cylinder. This press increased the rate of production four times.

The next major advance was the rotary press in which the traditional flat type-bed was abandoned and type and paper revolved together, each on separate cylinders, in continuous motion. The first successful rotary press was perfected by the American Richard M. Hoe in 1846. It was adopted in England by 1857 and improved nine years later by the replacement of sheets of paper with a continuous roll. By 1870 the machine was able to cut the printed roll and fold the newspapers. Meanwhile a variety of smaller presses was being developed for book

* The movable tray (or bed) carrying the type was pushed to and fro on rails and the tympan (frame) was introduced to hold the paper in position during printing.

228   *Automatic silk screen press, 1958*

227   *Rutt's hand-operated printing press. From Hansard's*
Typographia, *1825*

230   *Linotype composer. Model 74, 1957*

229   *Monotype standard keyboard, 1957*

and general printing.

The mechanization of typesetting came later and despite the introduction of a number of composing machines, these were not very satisfactory and traditional hand-setting, in which each letter had to be handled individually by the compositor, was still in force until the nineteenth century was well advanced. It was Ottmar Mergenthaler, a German emigré, who invented in America the first modern type-composing machine, suited especially to newspaper production, which was first used by the *New York Tribune* in 1886. The machine was called the Linotype (from line o' type) as it produced, by a continuous process, controlled by a keyboard operator, complete lines of metal type (230). In 1887 the American Tolbert Lanston invented his Monotype machine, also operated from a keyboard, which was suited to the printing of books (229).

Meanwhile new processes were being developed for printing illustrative matter in a faster and more varied manner. By the late Middle Ages finer, more delicate designs than before could be cut on hard-wood or metal blocks and in the sixteenth and seventeenth centuries were developed the crafts of copper engraving, drypoint, etching and mezzotint. But it was during the nineteenth century that first the iron printing press and later the powered presses gave sufficient pressure to print pictures with letterpress to illustrated books. Periodicals and newspapers were thus produced in great quantity.

By the 1880s photo-mechanical blocks were being made, but these were suited only to reproduce line drawings. The half-tone process, introduced about the same time, made it possible to reproduce a photograph or tone drawing, simulating the tones by a medium of thousands of dots of differing size. A process camera, in which a cross-line screen has been inserted in front of the negative plate, is used to copy the original. The light passing through the screen is broken up, so making the dots. Lithography, a process making use of the natural repulsion of water to grease, was invented in the 1790s by J.A. Senefelder of Prague. The intaglio process of photogravure was put forward by Fox Talbot as early as 1852, when he experimented with etching aquatint-screened plates with ferric chloride. Modern photogravure, using directly etched copper cylinders for printing, derives from the work of Karl Klic in the 1880s. Increasingly, these different printing processes have been adapted for colour work (228).

During the twentieth century, but particularly since the 1950s, printing methods have been revolutionized by the dominant role of photography combined with the use of electronics and computer technology (page 168). The advent of the offset lithographic press came in the first decade. Some time earlier metal plates had replaced stone for lithographic processes, but it was the introduction of the more resilient rubber 'blanket' cylinder which was novel. In this type of press the image was 'set-off' from the plate to the rubber and from thence transferred to the paper. Between 1910 and 1950 the offset press was made more efficient; electric motors were introduced to power it, new mechanical feeding devices were incorporated (web-fed offset presses), plate-making was improved, accuracy was better controlled and speed increased. Gradually the fast offset machine has taken over from the traditional letterpress printing model.

Similarly, since 1950, the new technology of photocomposition has revolutionized typesetting (230). Machines were devised which composed type on film. In these the keyboard operation produced a punched tape which incorporated all the type and layout data; the tape was then used to control the phototypesetting machine to imprint the film. During the 1960s and 1970s more advanced composing machines began to generate letters electronically and make use of computer memory systems and so lead to extremely fast typesetting.

THE TRANSMISSION AND REPRODUCTION OF SOUND

The telephone

By the 1870s experiments were being made in several countries to try to transmit the sound of the human voice by electrical means. The first practical instrument was devised by Alexander Graham Bell (1847–1922), a teacher of speech to the deaf (232), but more than one inventor contributed to the development of the telephone.

As early as 1861 a young physics teacher in Germany, Philipp Reis, devised an instrument to transmit (imperfectly), musical notes which he demonstrated in Frankfurt, but he died before it was perfected. In America Elisha Gray of the Western Electric Company was working on the idea in the 1870s, as also was Edison (page 119), but it was Bell who, on 10 March 1876, first transmitted speech by the instrument which he had invented.

The basis of telephony is to convert the sound waves of the speaker's voice into related electric waves which may be transmitted some distance, then transformed once more into sound waves to be received by the listener. Bell's 1876 instrument consisted of a transmitter in which a needle, attached to the centre of a parchment diaphragm, dipped into a container of acidulated water, vibrated to the sound of a human voice; it used a vibrating reed receiver. It was found that the vocal sounds were weak and long-distance communication unsuccessful until the invention in 1878 of an amplifier by David E. Hughes (1831–1900), a device which he called a microphone. This increased the sensitivity of the transmitter by using carbon rods as a suitable conductor. Later that year, in an instrument devised by Henry Hunnings, the bars were replaced by carbon granules inserted behind the diaphragm.

The possibilities of telephony were quickly appreciated. Bell designed improved instruments and demonstrated them in Philadelphia in 1877. The following year he visited England and demonstrated a model to Queen Victoria, who was most impressed. Telephone exchanges were set up to operate commercially, the first in Connecticut in 1878 then, in August 1879, in London. Other inventors designed more advanced telephone apparatus, Edison and Berliner in 1877 (pages 159, 160), followed by Hughes, then Hunnings.

Since 1900 the subscriber's telephone instrument has gradually become more convenient. The fundamental requirement remains: the speaker at each end of the line needs a transmitter and a receiver as well as a means of signalling to gain attention. For many years the speaking and listening instruments were separate, at first as part of a wall design (233), later a desk type (235).

Dials appeared as the automatic system was developed and in the USA in 1963 push buttons were introduced to replace dials (231).

Trunk lines were being laid by the end of the nineteeth century; Bell himself opened the first line from New York to Chicago in 1892. In Britain the overhead network had been replaced by the 1920s with underground cables and in 1980 British Telecom inaugurated the beginning of the changeover in Europe from the transmission of sound by radio waves to one by pulses of light produced by laser. The expensive heavy cables are being replaced by cheaper, finger-thick ones containing hair-thin optical glass fibres which will no longer be subject to electrical interference. Each of these optical cables can carry 100,000 two-way conversations (234).

Recording and reproduction: phonograph to music centre

Whereas the basis of telephony is the transmission of sound waves over a distance, in recording and reproduction the sound is preserved (by mechanical or electrical means) and reproduced at will. Thomas Alva Edison (1847–1931), the American inventor (page 119), was the first person to do this successfully on his talking machine (which he called a phonograph) of 1877, when he recorded his recitation of 'Mary had a little lamb' (236). His first machine was a tin-foil-covered brass drum about 100 mm in diameter on one side of which was fitted a recorder with mouthpiece, diaphragm and stylus and on the other a similarly designed reproducer. To record or reproduce sound the handle had to be turned. Some years passed before Edison's phonograph could be developed sufficiently for commercial sale. Then in 1895, fitted with reliable clockwork motors, a series of models was manufactured intended particularly for home use. These had wax cylinders (which played for two minutes) and the sound was amplified by a large horn (238).

Meanwhile other inventors were experimenting to try to produce a more realistic and long-lasting sound recording. The graphophone (239) was designed by C.S. Tainter (1854–1940) and C.A. Bell (1848–1924), the latter a cousin of the inventor of the telephone (page 156). The

231 *Modern press button (digital) telephone*

232 *Alexander Graham Bell*

23ᴬ *Jan Harrison is holding the new optical fibre telephone cable. Behind her is a drum of the older cable which it replaces*

233 *National Telephone Company wall telephone c.1910. Powered by battery at the exchange*

235 *Ericsson table telephone set c.1890. In general use for over 20 years*

237   *The Berliner gramophone of 1892. The discs are five inches in diameter and the handle has to be rotated at 150 revs. per minute*

236   *Thomas Alva Edison, 1911*

238   *Edison opera concert phonograph, 1912*

240   *Sony Music Centre HMK 5000, 1981*

239   *C.S. Tainter dictating to the graphophone*

instrument, which was marketed by Columbia, stemmed from Edison's phonograph and also had cylindrical records. But, in 1888, Emile Berliner demonstrated in Philadelphia his gramophone which recorded on flat discs in a spiral groove which had been etched continuously on the surface (237). He then invented a means of copying numbers of discs from a zinc master and from the 1890s onwards the disc record gradually became established as the pattern for twentieth-century recording, although phonographs continued to be made until the First World War. In 1898 the Gramophone Company was set up in London to market Berliner's instrument and by 1907 it was being manufactured in Hayes. In 1898 the 'dog model' had appeared which bore the famous pictorial trademark of 'His Master's Voice': it sold at £5–25.

During the 1920s the gramophone became a popular means of home entertainment. The instrument (including the horn amplifier) was then enclosed in a cabinet. Still hand-turned, it had doors to open and a lid which covered the turntable and playing arm. Wooden needles then gave place to fibre and later steel; sapphire, and now diamond, styluses (needles) are used in gramophone pick-ups. Sound recording, which up to then had been developed empirically, continued its progress scientifically. With the coming of electronic and vacuum technology the acoustic recording method, operated mechanically, was steadily supplanted in the 1930s by electrical recording and operation. In 1948 the long-playing microgroove record (L.P.), revolving at $33\frac{1}{3}$ revolutions per minute, was introduced and soon completely ousted the slower '78s'. Soon high-fidelity (hi-fi) equipment, designed to render as faithfully as possible the original sound being recorded and stereo two-channel systems of reproduction (available since 1954), were bringing a vivid, high quality sound into the living room.

The idea of reproducing sound by means of electromagnetism had been considered by both Edison and Tainter, but it was the Danish engineer Valdemar Poulsen (1869–1942) who first did this in 1898. Poulsen used steel wire for his recorder, but this was not very successful due partly to the material and to the lack of a suitable amplifier. It was the invention of the valve amplifier in 1930, the coming of radio and the development of the use of plastic magnetic tape in Germany during the Second World War that led to its widescale employment for modern cassette tapes. In this system, the signal from the amplified electric currents is fed into the recording head (an iron-core electromagnet) and a magnetic tape is drawn over it. In playback the action is reversed and, when the magnetized tape is passed over the playback head, the electric currents from the tape are amplified and reproduced. A modern music centre with stereo reproduction combines in one instrument a radio, the means of playing recorded discs and recording and playing magnetic tape cassettes (240).

Radio

Electromagnetic waves are used to send and receive information in both radio-telegraphy and radio-telephony; in the former case they are in the form of Morse code and, in the latter, human speech. The English physicist James Clerk Maxwell (1831–79) postulated in 1864 by mathematical theory the existence of electromagnetic radiation and he showed that light consists of electromagnetic waves which travel at 300 million metres per second (242). Over 20 years later the German physicist Heinrich Hertz (1857–94) confirmed Maxwell's theory by a series of experiments which he carried out between 1886 and 1889 at Karlsruhe (243).

During the 1890s a number of scientists in Europe and America were experimenting on the subject. Notable among these were Sir Oliver Lodge in Britain, Alexander Popov in Russia, Augusto Righi in Italy, Reginald Fessenden in America, Karl Braun in Germany and Edouard Branly in France. Each contributed new knowledge and invention to the sum of understanding, but it was the virtually unknown young Guglielmo Marconi (1874–1937) who demonstrated the workable possibilities and future commercial and social scope of wireless (radio) telegraphy. Marconi originated few new techniques himself but skilfully used what was already known and published. As he stated with some modesty, he 'experimented with devices invented by other men and applied certain improvements to these' (241).

In 1894 Marconi studied Hertz's contribution,

241   *Guglielmo Marconi*

242   *James Clerk Maxwell*

243   *Heinrich Hertz*

244   *Crystal set 1924–5 with earphones*

built himself a transmitter and, the following year at his home in Bologna, succeeded in transmitting signals over a range of two kilometres. He came to England in 1896 where he continued to experiment and transmit, work which culminated in the historic transmission of 12 December 1901 of three dots (Morse code for S) across the Atlantic Ocean from Cornwall to Newfoundland.

The invention in 1904 of the thermionic vacuum tube by Sir John Ambrose Fleming (1849–1945) was of vital importance in making possible the development of radio-telephony. Fleming, who had studied under Maxwell, was professor of electrical engineering at University College London (245). While there he designed this rectifying valve containing two electrodes (diode) which acted as a much more reliable detector of radio waves than devices which had been available previously. Two years later the American Lee de Forest improved the tube by adding a third electrode (so making it a triode) which would amplify and modulate as well as detect the radio waves. This eventually made it possible to manufacture highly sensitive wireless receivers and to pick up radio-telephone messages from much greater distances than before. The stage was set for the advent of radio broadcasting.

The American Reginald Fessenden (1866–1932) of Pittsburgh University first broadcast the human voice by means of radio waves in 1902. This was only over a distance of a mile, but four years later he successfully broadcast speech and music over 200 miles. Further experimental transmissions were made in Britain as well as the USA and wireless telephone sets (as they were called) were being manufactured but, before the First World War, radio telephony was thought of only as a medium for sending messages from individual to individual. During the war its military communication potential was partially realized and utilized by several countries, but after 1918 the possibilities of broadcasting for entertainment and commercial purposes were dawning. In Britain a transmitter was set up at Chelmsford which, by 1920, was broadcasting short daily programmes of news and music. Two years later the British Broadcasting Company,

then a business enterprise, began to transmit programmes from London under the call sign of 2LO. Expansion was rapid and wireless, as it was called until after 1945, was taken up by the public with such enormous enthusiasm that by 1927, when the British Broadcasting Corporation was granted its Royal Charter, over two million licences for receiving sets had been issued by the Post Office and a new, high-powered transmitter had been set up at Daventry (opened 27 July 1925).

In the early 1920s most receivers were crystal sets, often home-made to instructions published in the popular wireless magazines of the time. The crystal diodes were generally of galena or carborundum on which a suitable location was found by means of a steel or copper 'cat's whisker' (244). The crystal set contained no source of energy; the power for the headphones came from the transmitter intercepted by the listener's aerial which, it was recommended, should be as large and high as possible. Crystal receivers had a very limited range. They needed frequent adjustment, interference was often troublesome and the tall aerials were vulnerable to lightning. Loudspeakers, like those of the contemporary gramophones (page 160), were available by 1923 to replace headphones and valve receivers gradually took over from crystal sets.

The years 1927–39 were the golden ones for radio; by 1939 there were nine million licence holders (246). During the Second World War, with television transmission closed down, radio came into its own, keeping the British people in touch with vital events. After the war two new developments improved radio reception and convenience and initiated a new generation of miniature receivers suited to carrying about and fitting into cars and other equipment: these advances were VHF tuning and the transistor. Both have been instrumental in maintaining a new mass audience despite the incursion of television. The BBC began FM/VHF transmissions in 1955. They presented an impressive improvement in freedom from interference and in quality of sound. FM, which stands for frequency modulation, had been demonstrated in America in 1935 by Edwin Armstrong. This

245   *Sir J. Ambrose Fleming, 1934*

246   *Philips receiver type 634A, 1933*

247   *Murphy AM/FM receiver, model 242*

248   *'Pam' transistor receiver, type 710, 1956. First transistor receiver design to be made in Britain, price 30 guineas*

system lends itself particularly to transmission of signals over short distances. It makes use of VHF (very high frequency). Note that the shorter the wavelength the greater the frequency of the electromagnetic oscillations (247).

The invention of the transistor (page 169) in 1948–51 by the three Nobel prize-winning scientists at Bell Telephone Laboratories in America – Bardeen, Brattain and Shockley – revolutionized the design of electronic equipment and, in radio and television, replaced the thermionic vacuum tube as a means of controlling electric currents. A transistor is a small electronic amplifying device which is made from a tiny piece of silicon germanium, or other semiconductor material. In Britain the first transistor models of radio receivers were introduced in 1956. These had the advantage of very small size, they operated at a low voltage suited to battery use in portable equipment, they were highly efficient, were free from vibratory disturbance and lasted well (248).

VISUAL TRANSMISSION, RECORDING AND REPRODUCTION

Television

The broadcasting of visual images through space by electrical means to be received in the family home is a seeming miracle which has become accepted in the second half of the twentieth century as an integral part of life: arguably the advent of television has changed people's domestic living habits more than any other single technological development. Yet, although the fact of the availability of television has had to await our own time because before the First World War technology was not sufficiently advanced, attempts to send line-drawn pictures by an electric-telegraphy method began before 1850.

Willoughby Smith of Britain made the first study of photoconductivity in 1873 and discovered the light-sensitive properties of selenium, which becomes a better conductor of electricity when it is exposed to light. This led to suggestions from scientists in several countries for utilizing the discovery for the transmission of pictures by scanning an image in order to break it down into a sequence of tiny pictorial elements.

The most practical of these was an idea for a mechanical scanner put forward by the German scientist Paul Nipkow in 1884. He devised a cardboard disc, with small holes arranged spirally near its edge, which would be rotated so that a beam of light from a lamp shone through the holes at the scene to be transmitted. He used a battery-operated photoconductive (selenium) cell to transmit the scene in the form of electrical impulses.

An alternative line of research considered employing a cathode ray tube. Partially evacuated glass tubes containing two metal electrodes had been used by Sir William Crookes (1832–1919) as early as 1879, but it was the adaptation of the tube by the German physicist Karl Braun (1850–1918) in 1897 which pointed the way to its use in television. Braun made the stream of electrons issuing from the cathode in a vacuum visible by coating the wider end of the tube with phosphors (page 120). The Russian scientist Boris Rosing then suggested in 1907 that Braun's tube might be used to reproduce a picture at the receiver. Rosing, however, still used Nipkow's mechanical scanner at his transmitter, but soon afterwards the English scientist Alan Campbell-Swinton proposed the use of cathode ray tubes at both the receiving and transmitting ends. Neither the mechanical nor electrical system could be further developed with the technology available at the time and progress was soon interrupted by the First World War.

In the early 1920s determined efforts were made in several countries to make television work. One of the most energetic pioneers was the Scotsman John Logie Baird (1888–1946). Though not primarily a scientist or an inventor, Baird became intensely interested in the idea of television and, despite lack of financial means, pursued his aims with dogged determination. After two years' work he devised a transmitter in 1925 (249). It was a crude apparatus using a mechanical scanning method with a Nipkow disc, but in the small, flickering pictures tones were discernible and faces could be recognized. Baird successfully demonstrated his invention at the Royal Institution early in 1926. He improved his transmitting and receiving apparatus consider-

249 *Baird's experimental television transmitter of 1925*

250 *Bush television receiver, 1939*

251 *Emitron television transmitting tube, which operates on same principle as Zworykin's iconscope. Developed 1933–5 in EMI Research Laboratories*

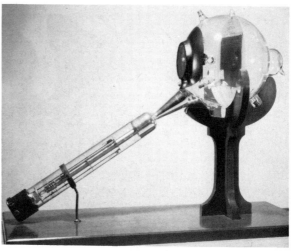

252 *Philips VR2020 video cassette recorder, 1981*

ably in the next few years and in 1929 the BBC agreed to begin an experimental service of five and a half hours a week transmission by the Baird Television Company. For this the Baird Televisor was marketed from 1930 at 25 guineas, but few were sold. The equipment now made use of the better light-sensitive photoelectric cells (page 164), but the fact that the system was still dependent on mechanical scanning of the Nipkow type meant that the pictures were poorly defined and it soon became clear that this method was leading to a dead end.

253   *Sony KV2201 television receiver. Trinitron colour*

This was chiefly because the alternative approach of electrical scanning, first propounded by Rosing and envisaged by Campbell-Swinton, had by this time developed sufficiently to provide pictures of much higher definition and quality. Much of the research for this achievement had taken place in the laboratories of the big American radio and gramophone companies. It was here that the electronics engineer Vladimir Zworykin (1889–1982), a Russian emigré who had

studied under Rosing in Leningrad, made a vital breakthrough by his invention of the iconoscope in 1923, which was perfected a few years later. This charge-storage transmitting tube was the first practical pickup (transmitting) device which could televise studio and outdoor scenes at a high standard of definition. In 1929 Zworykin also produced the first satisfactory tube for television reception.

In Britain J.D. McGee and his research colleagues at EMI laboratories were developing the emitron electronic television camera which was based on the iconoscope but was more sensitive (251). Subsequent research into television transmitting tubes has produced the orthicon (invented in the USA by Iams and Rose in 1939 as a useful advance on the iconoscope), the photoconductive tube known as the vidicon (developed in 1949 by Weimer, Forgue and Goodrich of RCA) and the Plumbicon introduced by the Philips Company in the Netherlands in the 1960s; this is a vidicon-type of tube with a lead-oxide target particularly suited to colour broadcasting.

On 2 November 1936 the BBC began transmitting from Alexandra Palace the first regular public television service in the world. They used, alternately, Baird and EMI systems, but the superiority of the EMI system, with its standard of 405 lines and 50 picture frames a second, soon became clear and in 1937 the Baird system was dropped (250).

In Britain television transmission closed down on 1 September 1939 for the duration of the war. It recommenced in June 1946 and during the 1950s became an increasingly important factor in home entertainment; in 1955 there were four and a half million licence holders. Expansion took place on all fronts: screens became larger, Eurovision exchange link-ups were made from 1954, commercial television began transmitting the following year.

Since 1955, when an experimental service was begun, colour television has gradually taken over from black and white, a process which has accelerated in the 1970s. There are two chief systems in operation commercially. In one the picture is formed by firing electron beams from three electron guns, one for each primary colour, through a perforated shield on to a screen containing about one million phosphor dots, colour-coded in groups of three, one red, one blue, one green. The beams react with the phosphors to produce the picture. In the other, one electron gun is used to produce three beams (253).

The launching of artificial earth satellites has made possible the world-wide live transmission of high-definition television. Earlybird was the first to be launched for commercial communication (1965) and it has been succeeded in the 1970s by larger, more complex and long-lasting satellites.

Viewdata and video-recording

Home television has now been extended by these two recent technological developments for information and entertainment. There are several information systems such as teletext (Ceefax and Oracle, for example) which broadcast their data on the television screen, as well as the more comprehensive viewdata systems, such as the British Telecom Prestel, which use the telephone network to link the user of an adapted television set to an immense store of computer-held information.

The development and marketing of video-cassette recorders is believed by many to represent an impending television revolution. This involves a visual extension of the magnetic tape audio-cassette (page 160) which makes it possible to record in the home any television programme so that it may be received at a different time from that transmitted. Systems such as Philips' VR2020 and Sony C5 and C7 enable several diffferent programmes on separate channels to be recorded during absences (252). Also, with the addition of a special camera, viewers may make and record their own programmes. A yet newer development is the videodisc system which incorporates hi-fi sound, but programme material for this has to be purchased.

CHAPTER NINE

Modern Technology: The Way Ahead

The mushroom growth since the Second World War of the electronics industry which, by way of the inventions of the transistor, the chip and the integrated circuit developed into microelectronics, has created what is often referred to as the 'second industrial revolution'. A revolution it certainly is, bringing in its train problems of industrial adjustment and unemployment but also untold benefits in eased, improved living and working conditions, transport, communications and entertainment. The first industrial revolution (pages 14, 15–20) relieved physical labour by mechanization; the second, by enabling a mechanical device to control itself as well as creating mechanisms which carry out the drudgery of calculation and considered judgement, relieved mental labour also.

Electronics deals with the design and use of devices which depend upon the conducting of electricity through a vacuum, a gas or a semiconductor. A semiconductor is a material halfway between a good conductor of electricity and a good insulator: metal is an excellent example of the former, mica of the latter. The conduction of electricity in a semiconductor is generally affected by the inclusion in it of certain impurities. Today, silicon is the most commonly used semiconductor material.

THE DIGITAL COMPUTER

Digital computers are not part of the furniture in the average home, but, during the last 30 years, their use has become so widespread that there are few areas of domestic life which are now not affected. Computers process and control, for instance, the borrowing of books from the local library, the handling of personal bank accounts, the automation of factory production of food, clothes, furniture. It is common practice to blame the computer when the gas bill registers £100,000 or when the bank account slides unaccountably into the red, but by now we all realize that it is human rather than mechanical error which is to blame.

A computer is an automatic device which performs calculations and processes information. In order to carry out this second task it has to be programmed, that is, instructed in a prescribed acceptable form (programming language) for the work which it is required to do. It stores this information in a 'memory' and can draw upon this in order to give answers to questions, evaluate courses of action or pronounce judgement upon alternative projects. This information usually appears in normal language on the screen of a cathode ray tube which is rather like a television receiver tube. The computer can manipulate immense quantities of information at very high speed. The computer system itself is referred to as 'hardware', the set of programs and data as 'software'. Computers have generally been designed to operate on a binary notation, that is, calculating by means of two numbers only, 0 and 1, instead of decimal (10) or duodecimal (12) systems which have traditionally been used in human calculation. Recently hexadecimal notation (employing 16 symbols) has been introduced, especially in microcomputing.

From the abacus of the ancient cultures to Napier's invention of logarithms (page 12) and on to the seventeeth-century adding machine of Blaise Pascal and the improved version made by von Leibniz, which multiplied by means of repeated addition, man has always sought mechanical aid to ease calculation. Punched card systems on the Jacquard principle (page 141) were introduced into American calculating machines by Hollerith in 1886, but it is the English professor of mathematics at Cambridge, Charles Babbage (1792–1871), who is credited with the concept of a computer, that is, an automatic machine to perform a sequence of different operations: to carry out at speed tedious and complex arithmetic, also to store this information for use in further calculation. Babbage spent much of his life perfecting the device which he designed in 1833, but it was never completed. It was operated on a punched-card system. He called it an analytical engine.

A century later in America development of the automatic computer was taken up once more and in 1945 the first electronic digital computer was built at the University of Pennsylvania. Replacing the earlier mechanical designs, this pointed the way to future advances; it was faster – it could perform 5,000 additions in a second – but it was cumbersome and weighed $29\frac{1}{2}$ tons. Storage of numbers and computing operations were handled by electronic circuits containing thermionic valves (18,000 of them) and punched cards or tape* were used for input and output data.

THE TRANSISTOR

It was not until the electric motor could be made small that its use proliferated, particularly in the field of domestic appliances (page 39). Similarly, the miniaturization of the electronic computer has rapidly and enormously increased its use. The first step towards this miniaturization of the computer was made possible by the invention in 1947 in the USA of the transistor (page 164). The

improved bipolar junction version of this tiny multielectronic semiconductor† device was developed by Shockley in 1949 (who also developed the alternative field-effect transistor) and ten years later it was well on the way to replacing the thermionic valve (page 162). Because of their small size transistors were used first instead of valves in hearing aids. By the mid-1950s they were in use for radio receivers (page 164) and, by the end of the decade, in computers. The replacement of thousands of valves (each about 100 mm long) in a computer by transistors (each about 5 mm long) made possible a tremendous reduction in the size of the equipment (257).

The use of a thyristor, another semiconductor device developed from the transistor, makes possible a more efficient control than previously of the development of electric power in electric motors and ovens. Thyristors have been incorporated for this purpose in such domestic appliances as, for instance, food processors and mixers (254).

THE INTEGRATED CIRCUIT AND MICROELECTRONICS

In the late 1940s the individual components of an electronic circuit had to be wired together. These components included thermionic valves (later transistors) to operate switching systems, resistors to resist the passage of electricity and capacitors to store and regulate the rate of production of an electric charge. The replacement of the thermionic valve by the transistor made the equipment much less bulky and more suited to the development of the printed circuit board. In this the circuit is printed on one side of an insulating panel and the components are wired up according to the circuit layout on the other side. Gradually components were further miniaturized and circuit panels became correspondingly smaller.

The miniaturizing process was then greatly advanced by Jack Kilby's achievement in America in 1957 of making the first effective semiconductor integrated circuit. In this, instead of components being contained and wired in-

---

*Apt, though dated, was Frank Muir's radio comment that 'computers have not eliminated red tape; they have only perforated it'.

† At first germanium was the chief semiconductor material, but this was later replaced by the more suitable silicon.

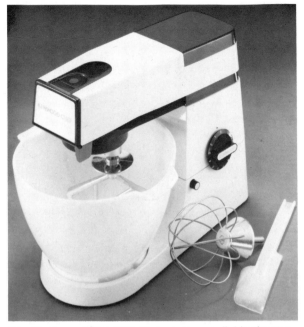

254   *The Kenwood Chef, 1978. Electronically controlled by thyristor*

255   *This photograph illustrates the smallness of the silicon microchip (also Lucy, see 256). These are normal no. 5 sewing needles, a pin and 40-gauge sewing cotton*

256   *Lucy (the SAA5070). The latest Mullard integrated circuit for teletext and viewdata systems, which contains its own microprocessor*

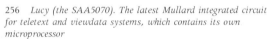

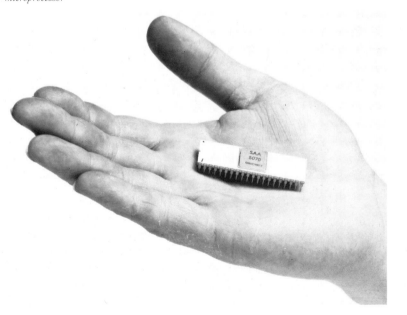

257   *(Left) A commonly used p-n-p bipolar junction transistor (type BC107). Three connector leads at base (Right) A commonly used miniature thermionic valve (type ECC88). Nine connector leads at base. Both drawings are actual size*

dividually, the whole electronic circuit is composed within one integrated container, a tiny wafer of silicon: the micro-chip. The transistor, resistor and capacitor are all based on the use of silicon. Since 1957 a whole new microelectronics industry has grown up, led by America, where so many laboratories and companies have been set up in the Santa Clara Valley south of San Francisco in California, that it has become known as Silicon Valley. The remainder of the industrialized world has followed the American lead.

The silicon integrated circuit possesses immense advantages over the former separately-wired component electronic circuits. Apart from the obvious one of miniaturization, it is far more reliable since the wiring was subject to human and material fallibility; also the time needed to assemble and wire the circuit has been saved. Chips are not cheap to make, for the process has to be precisely controlled under stringent conditions, but once made their use is legion and they can be easily and inexpensively built into a variety of equipment. Miniaturization has now reached an incredible level: the electronic heart of a computer, (the so-called 'central processor unit') which in 1950 occupied about a volume of two cubic feet, may now be accommodated on a silicon chip which is only five mm square (255, 256).

THE MICROPROCESSOR

This is the heart, the central processing unit of a computer system realized on a single silicon chip or a small group of chips. It is, in effect, an electronic calculator which may be programmed. It was the introduction of the microprocessor by the Intel Corporation of Silicon Valley in 1971 which initiated a rapid expansion of microelectronics. At first the microprocessor was used in pocket calculators and digital watches, but within the last two to three years its suitability to a much wider range of equipment has been envisaged.

Available at present in home appliances are cookers, toasters, washing machines, smoke detectors and sewing machines fitted with their own microelectronic programmers. Pre-set controls for cookers and washing machines have been available for some time, but control by

microprocessor is more sophisticated and reliable, it saves power, is easy to operate and provides a wider range of programs (259). In washing clothes the correct time and temperature for each fabric is precisely monitored; in cookers, both gas and electric, this is equally applicable and if the cook (unwisely or inadvertently) sets the timer incorrectly, the cooker will register 'error' on its digital display and refuse to carry out the instructions. The memory recall is also probably more accurate than that of the cook (258). Toasters make toast more consistently because, since the 'timer' works on temperature not time, no matter how many slices are cooked one after the other at a given setting, the toast will be the same golden brown (261). The new Singer Futura sewing machine has made sewing easier and more adventurous. The microprocessor provides an easy control of a wide range of stitches and operations, all stages being visually presented on a digital control and memory panel (262).

These examples are only the beginning and, because such new models will cost more, it may be some time before they are generally accepted. But possibilities for the near future include microelectronic control of the heating and lighting systems of the entire house to take account of the time of day and season, also differences of need on weekdays, Sundays and holiday-time. Such central control could handle many other functions, for instance switching on television or recording telephone calls. In cooking, temperature sensing probes could be inserted into the food so that, by using a temperature guide rather than time, the cooker would automatically compute the time needed to follow the housewife's instructions to 'cook the beef rare'.

In addition to specific household appliances, domestic life is also being changed by the incorporation of microelectronic systems on a wider front. Television games and viewdata presentation as well as remote switching control are already available and increasingly used (page 167). In printing computer control is now widely used in all fields of work, for instance, photo typesetting, inking systems, plate-making, page make-up and colour scanning. The decorative cutting machines making table glass may be

258   *Creda Carefree electronic cooker, 1981. Microprocessor control of clock and timer, six digital display panels, ceramic hob*

259   *Creda MicroElectronic 1000 Washing machine, 1981. Microprocessor program control*

260   *Philips microwave oven 500, 1981*

261   *Russell Hobbs electronic pop-up toaster. Model 5431, 1980. Microprocessor control with seven settings*

computerized. Congestion at the supermarket checkout may be relieved by electronic means: the checkout operator simply passes a wand over the bar code displayed on each article selected to register the purchase. These codes, which show parallel lines of varying thickness, identify the article concerned.

THE MICROWAVE OVEN
Microwave cooking (page 65) is a new, different method which makes use of radio waves of length as short as 12 cm. The electronic devices which were first used to generate these very short radio waves for radar purposes were developed in Britain and America during the Second World War. The first microwave cooker was produced soon afterwards in America and the method of cooking introduced into Britain in 1959.

In all forms of cooking the effect of the heat is to raise the temperature of the food and so increase the agitation of its molecules. In microwave cooking the very high frequency radio waves (microwaves) penetrate easily to the centre of the food and raise its temperature. So the heating effect is produced within the food itself and is not, as in conventional cooking, transferred to the surface of the food from an exterior source. The design of the inside of the microwave oven helps the concentration of the microwaves in the food by having reflecting polished interior surfaces and by the use of a rotating fan attached to the roof from the blades of which the microwaves are scattered. Microwave cooking will not 'brown' food; this must be done separately.

The chief advantage of microwave cooking is its speed: five minutes for a jacket potato. Also, food is defrosted in a fraction of the normal time: a medium-sized chicken in less than half an hour. Much less electricity is needed than in traditional cooking methods and, as it is only the food that is heated and the oven stays cool, spillage is easy to clean off. The output power of most microwave ovens is thyristor-controlled while in the USA microprocessor control is being introduced (260).

ELECTRONIC NEWS GATHERING
This system, known as ENG, began operating in British television in 1980 after a year's trial during 1978 and has also been introduced by some newspapers. For television, highly mobile electronic equipment is used – a camera, videotape recorder and a radio link – to transmit news directly from its location to the television studio. This may be done either by feeding the information over the Post Office cable network to the Television Centre or by 'beaming' it (in the London area) from a dish aerial on a Range Rover via a radio link to the top of the 400-feet high Millbank Tower and thence by cable. In both cases the system is much faster as both film processing and transportation time is saved (263). In the case of newspapers, the reporting of news stories by journalists is equally speeded up. Instead of telephoning to a typist at the newspaper office, as previously, the reporter types direct into portable computer terminals (about the size of a portable typewriter) information which is then rapidly transmitted to a central computer.

LASERS
The word laser is an acronym for light amplification by stimulated emission of radiation. It is a device which generates and amplifies optical waves in the visible part of the spectrum and produces a narrow beam of light of intense energy which may be pulsed or continuous. The first laser action was obtained in the USA in 1960 by Theodore Maiman, using a ruby crystal which emitted pulses of red light lasting 1/1,000th of a second. In the following year a continuously operating gas laser was developed at the Bell Telephone Laboratories in the USA. Since then a further range of lasers has been produced from various materials and the importance of the laser has grown as the range of uses to which it might be put has multiplied.

Lasers are extensively employed in industry for welding, fusing and cutting. They have become important tools in delicate surgery, notably detached retina of the eye, and fulfil varied needs in the space programme, the military field and in communications. Nearer to the domestic front the laser has become of considerable importance to the printing industry (page 156), in gravure technology, photo typesetting, photography and plate-making. Lasers are particularly used to transform digital inform-

262 *The Singer Futura 2001 electronic sewing machine. Stitch selection controlled by microprocessor*

263 *Electronic News Gathering (ENG) equipment being used by the BBC on location in Blackpool, 1978*

ation in computer storage into a printing image on a plate. Laser xerographic units are achieving work of excellent quality. British Telecom's introduction of optical glass fibres to replace the old cabling utilizes laser light waves (page 157). The employment of a low-power laser provides an alternative scanning method (page 173) at present being experimented with by the large chains at supermarket checkouts. It is expected that a wider range of uses for the laser beam in many areas of modern life will materialize.

Select Bibliography

AGIUS, P., *British Furniture 1880–1915*, Antique Collectors' Club, 1978.

ALTICK, R.D., *Victorian People and Ideas*, Dent, 1974.

ANTHONY, H.D., *Science and its Background*, Macmillan, 1971.

ASHLEY, M., *Life in Stuart England*, Batsford, 1967.

ASHTON, T.S., *The Industrial Revolution 1760–1830*, Oxford University Press, 1966.

BARLEY M.W., *The House and Home*, Studio Vista, 1971.

BERNAL, J.D., *Science in History*, Watts, 1965.

BONNETT, H., *Farming with Steam*, Shire, 1974.

BOWERS, B., *R.E.B. Crompton Pioneer Electrical Engineer*, HMSO, 1969.

BRIGGS, A., *The History of Broadcasting in the United Kingdom* (4 Vols.), Oxford University Press, 1961–79; *Victorian Cities*, Odhams Press, 1963; *The Nineteenth Century: the Contradictions of Progress* (Ed.) Thames and Hudson, 1970.

BRUTON, E., *Diamonds*, N.A.G. Press, 1978.

CAMPBELL, W.A., *The Chemical Industry*, Longman, 1971.

CAWOOD, C.L., *Vintage Tractors*, Shire, 1980.

CECIL, R., *Life in Edwardian England*, Batsford, 1969.

CHALONER, W.H., and MUSSON, A.E., *Industry and Technology*, Vista, 1963.

CHAMBERS, J.D., and MUNGAY, G.E., *The Agricultural Revolution 1750–1880*, Batsford, 1978.

CLAPHAM, J.H., *An Economic History of Britain 1820–1929* (3 Vols.), Cambridge University Press, 1950–2.

COOTES, R.J., *Britain since 1700*, Longman, 1970.

CORINA, M., *Pile it high, Sell it cheap; Biography of Sir John Cohen, Founder of Tesco*, Weidenfeld and Nicolson, 1971.

COURT, W.H.B., *A Concise Economic History of Britain, from 1750 to Recent Times*, Cambridge University Press, 1976.

CROWTHER, J.G., *British Scientists of the 19th Century*, Kegan Paul, Trench and Trubner, 1935; *British Scientists of the 20th Century*, Routledge and Kegan Paul, 1952; *Discoveries and Inventions of the 20th Century*, Routledge and Kegan Paul, 1966.

DAMPIER, W.C., *A History of Science*, Cambridge University Press, 1966.

DERRY, T.K., and WILLIAMS, T.I., *A Short History of Technology*, Clarendon Press, Oxford, 1960.

DIDEROT, D., *A Diderot Pictorial Encyclopaedia of Trades and Industry*, Dover Publications, New York, 1959.

ENCYCLOPAEDIA AMERICANA

ENCYCLOPAEDIA BRITANNICA

ENGELS, F., *The Condition of the Working Class in England* (English translation), Blackwell, 1971.

FEARN, J., *Thatch and Thatching*, Shire, 1976.

FISCHER, R.B., *Science, Man and Society*, Saunders, Philadelphia, 1971.

FORESTER, T., (Ed.) *The Microelectronics Revolution*, Blackwell, 1980.

GEDDES, K., *Broadcasting in Britain 1922–72*, HMSO, 1972.

GIEDION, S., *Mechanization Takes Command*, Oxford University Press, 1970.

GILBERT, K.R., *Early Machine Tools*, HMSO, 1975; *Henry Maudslay*, HMSO, 1971.

GOODMAN, J.W., (Ed.) *Laser Applications*, Academic Press, 1980.

HOOPER, M., *Everyday Inventions*, Angus and Robertson, 1976.

HUDSON, K., *Building Materials*, Longman, 1972; *Food, Clothes and Shelter*, John Baker, 1978.

JOY, E., *The Country Life Book of English Furniture*, Country Life, 1964; *English Furniture 1800–1851*, Sotheby Parke Bernet Publications, Ward Lock, 1977.

KRON, J., and SLESIN, S., *High-Tech: The Industrial Style and Source Book for the Home*, Allen Lane, 1979.

LAMBTON, L., *Temples of Convenience*, Gordon Fraser, 1978.

LARSON, E., *A History of Invention*, Phoenix House, 1961.

LAW, R.J., *The Steam Engine*, HMSO, 1977; *James Watt and the Separate Condenser*, HMSO, 1976.

LAYTON, C., *Ten Innovations*, Allen and Unwin, 1972.

MAYALL, W.H., *The Challenge of the Chip*, HMSO, 1980.

MESSHAM, S.E., *Gas: An Energy Industry*, HMSO, 1976.

MOODY, E., *Modern Furniture*, Studio Vista, 1966.

PYKE, M., *Food Science and Technology*, John Murray, 1970; *The Science Century*, John Murray, 1967.

QUENNELL, M., and C.H.B., *A History of Everyday Things in England* (5 Vols.), Batsford, 1976.

READER, W.J., *Life in Victorian England*, Batsford, 1967; *Metal Box: a History*, Heinemann, 1976.

REES, G., *St Michael: A History of Marks and Spencer*, Weidenfeld and Nicolson, 1969.

REEVE, R.M., *The Industrial Revolution 1750–1850*, University of London Press, 1971.

SEAMAN, L.C.B., *Life in Britain Between the Wars*, Batsford, 1970.

SINGER, C., and others (Ed.), *A History of Technology* (7 Vols.), Clarendon Press, Oxford, 1958–78.

SMITH, A.G.R., *Science and Society in the 16th and 17th Centuries*, Thames and Hudson, 1972.

STANLEY, C.C., *Highlights in the History of Concrete*, Cement and Concrete Association, 1979.

SUTCLIFFE A. (Ed.), *Multi-Storey Living: The British Working Class Experience*, Croom Helm, 1974.

TOLANSKY, S., *The History and Use of Diamonds*, Methuen, 1962.

WARNERS, A.W., and others (Ed.), *The Impact of Science and Technology*, Columbia University Press, 1965.

WHITE, R.B., *Prefabrication: a History of its Development in Britain*, HMSO, 1965.

WHITE, R.J., *Life in Regency England*, Batsford, 1967.

WILLIAMS, E.N., *Life in Georgian England*, Batsford, 1967.

WRIGHT, L., *Clean and Decent*, Routledge and Kegan Paul, 1960; *Home Fires Burning*, Routledge and Kegan Paul, 1964.

YARWOOD, D., *The English Home*, Batsford, 1979; *The British Kitchen*, Batsford, 1981.

YOUNG, E.C., *The New Penguin Dictionary of Electronics*, Penguin Books, 1979.

Index

Illustration references are printed in **bold type**